Lew McCoy On Antennas
Pull Up A Chair and Learn From The Master

Lew McCoy On Antennas
Pull Up A Chair and Learn From The Master

By Lew McCoy, W1ICP

CQ Communications, Inc.

Second Printing 1997
Library of Congress Catalog Card Number 94-69519
ISBN 0-943016-08-8

Editor: Gail M. Schieber
Technical Advisor: Lew Ozimek, N2OZ
Layout and Design: Elizabeth Ryan
Illustrations: Hal Keith, K&S Graphics
Cover photo: Larry Mulvehill, WB2ZPI

Published by CQ Communications, Inc.
76 North Broadway
Hicksville, New York 11801 USA

Printed in the United States of America.

Acknowledgements

In producing a book, usually many people besides the author are involved. When one has been a writer for as many years as I have, one tends to become overconfident. In many cases, the author cannot see the forest for the trees, if you'll forgive a tired cliche. I certainly owe special thanks to Gail Schieber, the Managing Editor of *CQ* magazine, who has kept me on the straight and narrow. Also, much thanks goes to Lew Ozimek, N2OZ, who helped make the book what it is.

Over the years I have worked with many exceptional people. Two of the finest are Dick Ross, K2MGA, Publisher of *CQ* magazine, and Alan Dorhoffer, K2EEK, Editor of *CQ*. These two really fine gentlemen are radio amateurs entirely dedicated to our hobby, and it has been a real joy to be closely associated with them.

A special note of appreciation and thanks must go to my long-suffering wife of nearly 56 years, who has put up with my quirks for all these years and who has never lost her ability to support and encourage me.

Last, a great deal of thanks must go to you, the readers who have accepted my verbal offerings over the years. Your constructive criticism, suggestions, and encouragement have always been welcome and certainly helped me to keep striving for better ideas and better ways of presenting them.

Lew McCoy, W1ICP
November 1994

Preface

When I was a young Novice in the early 1950s, every month I anxiously awaited the arrival of *QST* and *CQ*. I poured through every issue for the latest information on technical topics, building projects, surplus, antennas, and all the other aspects of amateur radio that enthralled me at that age. When I was slightly older and able to get to local hamfests, I remember meeting some of the people whose articles I had read and whose designs I had diligently copied. I really felt that I was in the presence of greatness. After all, what does a kid know?

I met Lew McCoy during this time, not directly, but through his articles and most importantly through his lectures and talks at hamfests. During all of these ensuing years I can't tell you the number of times I've listened to him speak on open-wire, antenna tuners, antennas, and TVI. It's one of those things you could sit through a million times and never get tired or bored. No, it's not that the concept of open-wire feeders is that spell-binding or even has the remotest possibility of being the next block-buster movie. It's just the way he spoke and still speaks about it.

In the early '60s when I went to work for *CQ* and began my odyssey through the hamfest circuit, I had the opportunity to spend more and more time with Mac, albeit the junior confrere. You could always ask a question, and invariably he had the answer. Early on it was apparent that he easily could frame the answer to suit the level of the questioner's understanding. The main thing he expressed then, and still does today, is a complete respect for the questioner or audience, giving everyone equal importance.

I think that fundamentally Mac is a teacher at heart. Those of us who have been around a long time have enjoyed his brand of education and wish we had been around longer to sample more. He's always been quick to champion a new writer or anyone with a good idea that he can include in both his writings and his talks. He still kids me about calling him "The Living Legend" when he came onboard *CQ*, but in a sense that's exactly what he is.

When the time came to discuss writing this book it became clear that what we all wanted was more than just a dry text full of facts and figures. It's hard to describe in words precisely what the goal was, but it had more to do with creating an atmosphere or living tapestry of one of our most important topics as amateurs—antennas. So if you're new to the hobby or have been around for a long time, what you're about to experience is in a sense a living book.

Just imagine your living room with a couple of easy chairs and a table to hold some cold drinks and a few snacks. You and Mac walk in, sit down, get comfortable, take a sip of your drinks, and begin to talk about antennas. It's more listening than talking. Oh, you may have a question now and then, but he seems to fit the answer right in and keeps on going. You do get some basic theory, basic in the sense that the terms take on meaning rather than being tossed around simply as labels. Need some examples? Well, here's what so-and-so is doing or has done, for example, and here's how it works. He produces some pictures and drawings to illustrate the point. For a few hours you have a famous, charming house guest sharing a lifetime of knowledge with you.

If you're at all familiar with his writings or his talks, that's the feeling you'll come away with—something very personal. In fact, you can almost hear his voice as you read. If you haven't read much of his work, you'll finish this book with a new friend. And if you happen to see him at a hamfest, he'll expect you to come over and say hello. I think it's also safe to say that it wouldn't take too much coaxing to get him to autograph his book.

After 40 years of reading Mac's work and listening to him speak, I'm glad that we at *CQ* are able to publish his first book on amateur radio. I don't know why it took so long for someone to talk him into it, but I feel fortunate that it was us.

Alan M. Dorhoffer, K2EEK
Editor, CQ Magazine
November 1994

Introduction

Every book should start out with an introduction, and that introduction should include background information about the author. I also feel that the author's qualifications and experience should be included so the reader knows he isn't getting a snow job!

I was born in Chicago in 1916, and my interest in amateur radio goes back a very long way. My father was a ham, operating a spark set back in the very early 1920s. The earliest memory I can recall is that of the roar and the sparks from the rotary gap coil shooting out streams of light as my father keyed it. Like many sons, however, I didn't take much interest in my father's hobby until many years later.

Shortly before World War II, I married and became a proud father. A short time later we moved to Blue Island, Illinois and bought a house. There I found I had an amateur for a neighbor, Roy Black, W9CYW, who was kind enough to show me his station and demonstrate how it worked. This probably was the real start of my interest in amateur radio.

About a mile away from our house I noticed another radio amateur's house with a home-built wooden tower next to it, upon which stood a homemade beam antenna (all beams were homemade in those days).

This piqued my curiosity. I summoned up enough courage to knock on this individual's door and asked if I could see his station. I was timid because this amateur, Bob Adair, W9RRX, was a big-gun DXer who I knew contacted stations all over the world and was thought to be inaccessible to ordinary humans. However, little did I realize at the time that all DXers are accessible and eager to brag about their DX accomplishments. In any case, I was welcomed and given a tour of his shack, which consisted of a National HRO receiver with several stages of preselection and a transmitter which used a pair of completely exposed 250 THs. For the neophyte, 250 THs are relatively large transmitting tubes which require very high voltages for operation, and by high, I mean thousands of volts.

Bob turned on his rig and immediately proceeded to shock me out of my socks by taking out a cigarette, leaning close to these tubes, and drawing a huge RF arc from the plate caps with a screwdriver to light his cigarette! As we say, those were the days. In any case, Bob became my "Elmer," or teacher, and led me into the fascinating mysteries of amateur radio.

During World War II amateur radio operation was very limited. After the war 10 meters was the first

Shown here are some of the antennas in my own station. W1ICP is in a constant state of flux, both as to station equipment and antennas. In this case, on the tower is the DJ2UT beam and in the foreground, mounted above my roof, is a Lightning Bolt quad I was testing at the time.

This photo should really make your mouth water. These are just a few of the antennas of the top contest station of Don Doughty, W6EEN. On the left is a full-size, four-element 40 meter beam. On the right is a multi-element 20 meter beam. Don has his towers oriented so that he can erect some really high 80 and 160 meter antennas.

band returned to us. I studied Morse code using a tape Instructograph machine. I also poured over the good old American Radio Relay League (ARRL) license manual to try to master the questions and answers. Eventually I took the Class B exam and was licensed as W9FHZ (Fanny's Handy Zipper). Now I was ready to set up a station.

An amateur who had a welding machine helped me make a three-element 10 meter beam consisting of a director, driven element, and reflector. The elements were thin-wall electrician's tubing.

While the antenna was being constructed, I built a three-stage transmitter. It was crystal controlled, with a crystal on 80 which doubled to 40, then doubled in another stage to 20, and then doubled again to 10. I assumed that I was all set to get on the air, so I turned on my rig and proceeded to completely fill up a log book with calls to other stations—with no answers. Talk about despair and frustration! I was not aware that my transmitter had a big problem. I finally called my friend Bob Adair and asked for his help. He came over with what he called a "tune-up loop," which was simply a loop of wire with a flashlight bulb connected in series. When this loop was held over any coil in question, the RF coupled to the loop made the light come on, indicating that there was power in the circuit.

You guessed it: The light on my middle doubler cir-

cuit didn't light. The middle stage apparently was tripling, putting my output signal on some unknown frequency. Bob thought for about two minutes and said, "This coil has too many turns of wire. Take off three or four turns." He stood by to confirm my modification and to check performance. I turned on the rig and listened to see if anyone was on. I found HZ1AB, a very rare station in Saudi Arabia, calling CQ, so I gave him a call. Meanwhile Bob went hell-bent out the door heading for home. The station in Arabia came back to me and gave me an S9+ report. It was about a mile to Bob's house, but he must have driven like a demon, because he suddenly broke in on the frequency and called the Arabian station. My report was S9+ with my 50 watts, and Bob got S7 with his kilowatt and beam!

What was the difference? Obviously, I had a much better location, or a better beam! We eventually determined that it was the beam, not the location. So there I was running about 50 watts, whereas Bob with his 1000 watts got a poorer report. The only difference was the effectiveness of the antenna. That certainly was an early lesson in the importance of antennas for good communication.

At that time in amateur radio very few of us were using coaxial feed lines. We primarily used open wire to a T matching section on the driven element. Beams

A close-up of the full-size 40 meter beam at W6EEN's QTH. Note the truss supports attached to this very large antenna, keeping the boom from sagging.

Another of W6EEN's antennas, but this one is no longer in use. It is a log periodic beam designed to give frequency coverage from 10 through 20 meters at gains of 6 through 8 decibels. According to Don, however, the beam was a real disappointment when compared to a 20 meter beam, being down as much as two S units.

were really new, and there was much arguing going on relative to element spacing, gain, and so on. Ed Tilton, W1HDQ, VHF Editor of *QST* magazine, and later when I went to work at the ARRL a very good friend of mine, had done considerable experimenting with elements spaced wider than was customary, with excellent results. My beam was one with wider spacing based on Ed's work, while Bob's was not. It should be said that we climbed Bob's wooden tower and quickly changed his beam—hi!

So those were my early days. I later moved to the Lake of The Ozarks in Missouri to try to get away from TVI problems. I became WØICP, but only stayed there for a little over a year. The ARRL was in a phone/CW war at the time, and their board of directors decided they needed a "phone" type to provide guidance to members. Since I was gradually starving to death in Missouri, I applied and got the job as Assistant Communications Manager in Charge of Phone Activities. I have grown more mellow over the years, but at that time a phone man was about as welcome at ARRL Headquarters as the plague. Despite my job description at the ARRL, my first assignment was the responsibility for code practice from W1AW!

However, at least I have the distinct pleasure of being the founder of the popular Novice Roundup and 10-Meter WAS contests, although nobody at the ARRL these days remembers that.

When the Novice Class license came into being (and I had a great deal to do with that), my mentor, George Grammer, W1DF, put me in charge of writing an article every month for Novice amateurs. He always said, "Lew, you don't know much, so you should be ideal to teach people who don't know much!" I never forgot the following important point, however: Antennas and location are the answers to questions relating to how to have an outstanding signal. My fascination with amateur radio therefore was not so much with transmitters and receivers, even though I built and described many, but rather it was with antennas.

The Technical Department of the ARRL at that time was in desperate need of someone to go out on the road and convince the public that TVI was not the fault of amateur operators. In my dim, dark past I had been a professional entertainer (magician), so I apparently was the ideal guy for the job. (Believe me, it took a magician to convince people we were not the cause of TVI!) I traveled the country in the early '50s lecturing on TVI and as a result became fairly well known.

Over long and happy years I have come to know some very smart—brilliant, really—people in radio, many of whom were very savvy antenna men. Without a doubt, three really stand out. One is George Grammer, W1DF, who was the technical editor of QST and the technical director of the ARRL. Another is Byron Goodman, W1DX, who was one of the real factors in making single sideband what it is today. Last, and by no means least, is Don Mix, W1TS, who was a great transmitter man. These three men advised me, encouraged me, edited my articles, and taught me about antennas. I owe a great deal to them. One last thing, which speaks for itself: The very famous *ARRL Antenna Handbook* was first put together by George Grammer and Byron Goodman!

In any case, I resided at a well of knowledge, and I profited from it. I quickly learned, for example, that antennas were and are important to all amateurs whether experienced old hands or novices. I studied and learned, and in my usual modest manner I will say that I think I became quite knowledgeable on the subject.

When you learn as I did, you quickly find out that there are some things that work and some that don't. There are some antennas that work, and of course some that don't. In this book the emphasis will be on antennas that I personally have built or used and that I know do a great job. By discussing my experiences, I will try to make you a very astute antenna person. Don't bypass the basics; learn them. Once you do, you will quickly see and understand what is good and what is not.

I hope you enjoy the book. Write to me if you like it. I also hope it helps you improve your capabilities, increase your enjoyment of amateur radio, and enhance your performance on the air.

Lew McCoy, W1ICP

Table of Contents

Chapter 1

Some Basic Facts About Antennas

Before discussing different kinds of antennas and their capabilities, I feel I should clarify a couple of points. The primary approach I will take in this book is to make my explanations as simple as possible, while still keeping them technically sound. For example, I could start out by explaining voltage and current distribution on wires with many charts and graphs, but it is very difficult to set down such information in very easy-to-understand language. This book therefore will follow a different path. If you want a much more technical discussion of antennas, then I recommend *The ARRL Antenna Book.* If you read and *understand* that book, you definitely will become an antenna expert. Or at the very least, when you tell someone you are an "extra" class amateur, you really will be one.

I spent several years as a contributing editor to *The ARRL Antenna Book,* and I have always had a desire to write an understandable book for the amateur. I feel that I have accomplished that here. The majority of amateurs are a long way from being technical engineers, and I have made every attempt to make this book understandable and useful. There are some simple, basic facts about antennas that will help the reader become more knowledgeable.

Something else I hope to accomplish in this text is a real-life comparison of multiband directional beam antennas. I do not recall ever having seen such a comparison. It is easy to read all the material presented by antenna manufacturers and get lost in the different claims. While I have lectured on the subject of multiband directional beam antennas many times, I have written about it only once. This book gives me an opportunity to cover that information once again.

In addition, at various times I will mention specific manufacturers of antennas and related items. I mention these in this book because I have had personal experience with these particular products. This does not mean there are not other good antennas, etc., out there. It just means that I personally have not had any experience with them. You will not find the manufacturers' phone numbers or addresses, or pricing information, simply because this information changes quite frequently. The reader can refer to *CQ* magazine or the most recent issue of the *CQ Equipment Buyer's Guide* for those specifics, and for the names and particulars of other dealers and manufacturers of products discussed in this book.

One other small point: I will repeat some items several times simply to drive home the point. Forgive me for such repetition, but it may help.

Antenna Feed Point

The feed point of all antennas is the point where the feed line is attached to the antenna to deliver radio energy. This feed point has certain characteristics which depend on many different factors. Fortunately, these factors are easy to understand. The key characteristic of a feed point can be defined by a single term—*impedance.*

The impedance of an antenna is always made up of at least two (and usually a third) properties. Because an antenna customarily is made of metal, the antenna will always have some actual ohmic resistance. In ohmic resistance, power is dissipated as heat and is lost. The power radiates, but only as heat. This resistance performs just as a resistor does. Another property of the feed point is radiation resistance. Unlike ohmic resistance, this is *called* a resistance, but it is that characteristic which determines what portion of the signal is coupled to space and radiated. Last, another "non-real" resistance exists which is called *reactance*. By calling reactance non-real, we are differentiating between this resistance and ohmic resistance, which is very real. Reactance is best likened to a gate or switch. If reactance is present, the switch is partially closed, preventing the flow of power into the antenna. Reactance is *only* present when the antenna is non-resonant.

Trying to illustrate the feed point of an antenna—to present in simple terms the ideas of radiation resistance, ohmic resistance, and reactance—is well nigh impossible. Instead of just glossing over this point, I am going to attempt to make these points clearer. I want you, the reader, to use your imagination.

Let's assume we have a water-tight metal box, which is going to represent the antenna. The impedance of this antenna/box is going to be defined in the same way as previously presented—namely by radiation resistance, ohmic resistance, and reactance. Now let's assume we connect a pipe (feed line) to this box and we flow liquid (RF energy) through this line from a source (transmitter). Around the box a series of holes will permit the power, entering the box, to flow or escape. We will call this escaping flow the *radiated RF*. If there are enough holes to spray out this power as fast as it enters the box, we can say the box is 100 percent efficient, because all of the power is immediately dispersed. We can put a value on this rate of spray radiation and call it *radiation resistance, or Rr.* Suppose, however, that there are not enough holes in this box to permit all of the power to escape instantly. When this occurs, the box is no longer 100 percent efficient. Assuming 95 percent efficiency, then 5 percent of the power cannot get out.

There is a direct similarity between this description and the way an antenna works. Now add one more factor—a valve directly at the entrance to the box. If the valve is partially closed, the flow of RF into the box is impeded. If it is wide open, it has no effect. Such a valve can be called *reactance* (and I am really trying to oversimplify here).

Let's carry this illustration a little further. Assume that the lower the frequencies used, the larger the size of the box must be. On 80 and 40 meters, for example, the largest box is used on 80, and it is twice the size of the 40 meter box. Continuing in this vein, 40 will be larger than 20, and so on. If we assume that the liquid, or RF, flow in our pipe for each band is different, such as being thicker or heavier for the lower band and thinner or lighter for the higher bands, then we can take our illustration even further.

In order to get the highest efficiency, or the most RF flow from the inside of the box to the outside of the box, we naturally need a larger box. Let's say we use a 40 meter size box for 80 meter RF. Efficiency would drop to somewhere around 50 percent simply because it would be difficult to transfer the 80 meter power as quickly from the inside to the outside of the box. So the smaller the box, the lower the efficiency. This concept is the same with an actual antenna.

Now that we have started thinking in the right direction, let's move to actual antennas.

Dipole Antennas and Impedance

A very common antenna in amateur radio is a dipole, which consists of two equal lengths of wire, or other metal conductors. Such a dipole is customarily thought of as two conductors with the overall length of both conductors equal to a one-half wavelength in electrical length. However, the conductors can be any length and still be considered a dipole.

The most common dipole used on the amateur 80 meter band is one-half wavelength long, or approximately 130 feet. While the dipole is cut for just the one band, it will work, and quite well, on all the other bands if it is tuned as a system. Many, if not most, amateurs use this particular dipole on more than one band, and we will cover that subject in detail later. The formula to calculate the length in feet of wire dipoles for a half wavelength is 468 divided by the frequency in megaHertz. For example, an antenna for 80 meters—say, 3.7 MHz—would be 468 divided by 3.7, or 468/3.7, which equals 126.48 feet. Simple enough? This brings up one small point that certainly bothers newcomers.

I recall the first dipole I built from *The ARRL Antenna Book*. Its length, from the formula, was listed as 127 feet 6 1/2 inches. Naturally, when making a dipole to suspend in the air, insulators must be used at the ends to support the antenna. Each half of the dipole was one half the dimension given above, or 68 feet 7/8 inch long. Nowhere in the book did it say how important it was to have or not have the exact length. Did the wire wrapped around the end insulators figure into this length? How important was it to be exact? The book simply didn't say.

I remember consulting several old timers, and they were universal in their statements that the length was not that critical and to forget about the few inches at the ends. They were right, in a sense. The exact length is certainly not critical on 160, 80, or 40 meters; a few

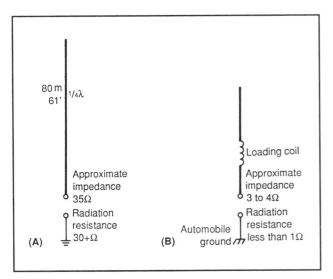

Figure 1-1. At the left is the normal 80 meter, one-quarter wavelength vertical. The impedance of this vertical is on the order of 33 ohms, depending on ground conditions. Of this 33 ohms, 90 percent is the radiation resistance, or useful portion, of the impedance. At (B) is a typical 8 foot, coil-loaded 80 meter vertical whip. In this case, because of the reduced size (it is still resonant) the impedance drops to just a few ohms, requiring matching to get to 50 ohm feed. However, the radiation resistance is going to be only a fraction of an ohm. That is why reducing ohmic losses by extensive vehicle bonding and grounding is so important in mobile work on the low bands.

inches give or take just isn't that important. On VHF and UHF, of course, fractions of inches can be important in antenna lengths, particularly when considering parasitic elements and their dimensions. However, on the low bands the height of the antenna above ground can have more effect on antenna impedances and resonances than the fractions of inches in lengths.

Getting back to the feed point or impedance, the normal wire dipole for 80 meters will have an ohmic resistance of 2 to 3 ohms. Most of this comes from the resistance in the wire itself. Actually, some of the resistance gets into the act because of resistance of insulators and the proximity of the dipole to nearby objects—trees, rain gutters, house wiring, power lines, and so on. Therefore, ohmic resistance is the first of the two common properties in the feed point.

The second element, while not a real resistance, is classified as such, and this is one element that can be confusing to newcomers. It is called *radiation resistance*, and its designation in formulas is Rr. An antenna can be said to be coupled to space, and the power fed to it will be coupled from the antenna and radiated into space. It is radiation resistance that determines

the efficiency of an antenna. The ordinary half-wave dipole can be said to be one of the most efficient antennas for a very simple reason. Assuming the antenna is at least one-half wavelength above earth, the impedance, or the sum of the two resistances, is going to be approximately 70 ohms. The ohmic resistance will be about 2 to 3 ohms, assuming No. 14 or larger wire, while the radiation resistance will be 67 or 68 ohms. This in turn means that if we fed 70 watts to the antenna, all but two or three watts would be radiated, making it a very efficient antenna indeed.

Repeating what we said earlier, the last property in the feed point is also expressed as a resistance, even though it is not a real resistance. It is called *reactance*. When an antenna is resonant—that is, its length equals the formula electrical length—only ohmic and radiation resistance are present. Only when the antenna is not resonant is reactance present. If an antenna is too long for a given frequency, it is said to have "inductive" reactance. If it is too short, the reactance is "capacitive." Power cannot be lost in reactance in the same way it is lost in ohmic resistance, but reactance, explained simply, is a property that stops or impedes the flow of power into the circuit—or antenna. It is simple enough to correct for reactance. Introduce an exact *opposite* amount of reactance and one cancels the other out, leaving just ohmic and radiation resistance.

Factors Affecting Impedance

What else controls the impedance of an antenna? Actually, there are several things, the most important of which is the antenna's height above perfect ground. A half-wave dipole exactly one-quarter or exactly one-half wavelength above such a ground would have a radiation resistance of 70 ohms. Below one-quarter wave the radiation resistance would decrease, until at ground level it would be practically zero. Above one-half wavelength the curve would level off and stay near 70 ohms. Keep this in mind, because when we talk in terms of antenna patterns it becomes important.

The length of an antenna has a significant control over the impedance of any antenna. There is one simple rule governing antenna impedance and that is, the smaller an antenna is for a given frequency, the lower the radiation resistance (Rr) will be. In turn, this affects the efficiency of the antenna. The most startling example of this is an 80 meter mobile whip antenna, usually on the order of 8 feet in length. The correct *electrical* length is usually achieved by using "loading" coils in the 8 foot whip. Figure 1-1 illustrates what I am writing about. A full-size quarter-wavelength vertical would be approximately 64 feet high. Now visualize a whip with a coil about 3 feet up from the bottom. About 4 feet of whip extends above the coil. We now have an 8 foot whip that is essentially a resonant *electrical* length on 80 (the resonant length is achieved because of the loading coil at the center), but it is only 8 feet long overall.

While we can resonate this "loaded" 8 foot whip to 80 meters, the antenna has a very low radiation resistance (recall that the radiation resistance is the useful part of an antenna's impedance). The Rr of this antenna is only a small fraction of an ohm. The ohmic resistance of this whip, depending on the grounding of the metal parts of an automobile, will run 3 or more ohms. Let's suppose that this whip has an impedance of 3.1 ohms. Three ohms is ohmic resistance and 0.1 ohm is radiation resistance. What happens if we now feed 310 watts to this antenna? That's right, 300 watts is dissipated as heat from the 3 ohms of ohmic resistance, while only 10 watts will be radiated! (Actually, the amount of power radiated is probably much less.) Keep in mind, therefore, that the smaller an antenna is physically for a given frequency, the more loss the antenna will have. I was taught, and very strictly I might add, that small antennas are inefficient, so avoid them if possible.

Running antennas near wire objects or any type of conductor can also alter the impedance. For example, we all are familiar with beam antennas with reflectors or directors. The impedance of the driven element by itself would be 70 ohms or so, but adding a director or reflector reduces this impedance to a lower value. This leads us to the subject of feeding RF energy to the antenna and feed-line matching to the antenna.

Chapter 2

Standing Wave Ratio (SWR) and Matching

There is probably no subject you are going to hear discussed and cussed more than standing wave ratio (SWR). Once you understand all the ramifications of SWR, you will begin to feel comfortable with the subject. There are many areas to cover, but none of them are really complicated.

SWR Defined

We have already discussed, and hopefully you have begun to understand, feed-point antenna impedance. Let's now attach a feed line to the antenna and see what happens. First, however, I will define SWR.

Radio frequency energy in the form of voltage and current flows along the feed line to the antenna. The standing wave ratio (SWR) is the ratio of either the maximum to minimum voltage or maximum to minimum current that exists at any given point on a transmission line. The objective in feeding antennas is to reduce the SWR to as low a ratio as possible, the ideal being 1 to 1. This is accomplished by matching the output of the transmitter to the feed line and the antenna.

Feeding Antennas

In the early days there was a lot of mystery about feeding antennas. This is no longer true. With modern computer programs you can determine the feed impedance of practically any antenna.

In the old days the feed line used to connect the amplifier to the antenna was usually an open-line variety—two identical conductors equally spaced to keep the two wires parallel and symmetrical. Such lines became known as "balanced" feed lines. Coaxial line also has two conductors—an inner plus an outer

conductor (shield) which surrounds the inner conductor, both conductors separated by a dielectric material. While coax is a symmetrical line, it commonly is referred to by amateurs as an unbalanced feed line.

Early Transmitter Amplifiers

Transmitter amplifiers of those early days were customarily designed for what was known as "link coupling" to connect the amplifier tank circuit to the open-wire feed line (figure 2-1). There were many tank-circuit designs, but believe it or not, few (if any) used the circuit that is common today. Nearly all tank circuits in current use are some form of the pi network originally designed by Art Collins, founder of Collins Radio. The reason for the change is simple enough, but the history is not.

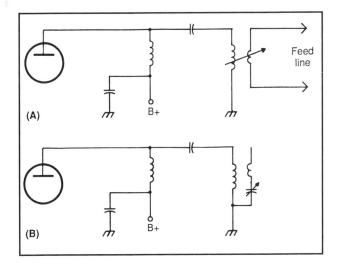

Figure 2-1. Illustrated at (A) is one of the common methods of using link coupling from the amplifier to the feed line. At (B) is a method used to adjust the link using a series variable capacitor for coax lines.

With early transmitters in order to change bands you had to change the coils in the transmitter using plug-in coils (or have individual amplifiers for each band). When World War II came along, the groundwork was laid for many basic changes in our hobby.

Feed Lines

Open-wire type line was inexpensive, easy to construct, and extremely efficient. In fact, it is still the most efficient line you can use. During World War II, however, coaxial feed lines, which were difficult if not impossible for amateurs to make, became extremely common. After the war this same coax entered the surplus market, where it was readily available and inexpensive. Coax was and still is popular because it is a shielded line, and unlike open wire it can be installed directly alongside metal objects or even buried underground without harmful effects. During the war, though, amateurs did not make a mass move towards coax simply because it meant a considerable redesign of the circuits in use. One more thing was then to happen to change our methods—and I might add forevermore.

The Impact of Television

In the late 1940s television suddenly became readily available. In the beginning TV sets were very expensive and usually only could be seen in local bars where they were used to attract customers—and they had 5 inch screens! Whenever an amateur station was operated near a TV set, the amateur transmitter created interference which either destroyed the picture or made it unusable. However, because TV sets were few in number then, amateurs were not really concerned.

I had seen television in the taverns (where I rarely went), but no one had a set anywhere near where I lived. That fall I went east to visit my wife's family, and when I returned two weeks later, I noticed a strange antenna on my next-door neighbor's house. You guessed it! He was the first one in a very wide area to purchase a TV set, and he had to be next door to me! He was a good friend and neighbor, but I have to admit, not for long. I probably can safely say that shortly after that occurrence I learned as much about television interference (TVI), public relations, and so on, as any amateur alive.

George Grammer, W1DF, Technical Editor of *QST*, and Phil Rand, W1DBM, as well as others, had determined that the only answer to TVI was complete RF

tight shielding of a transmitter and the use of low-pass filters on the transmitter as well as high-pass filters on the TV sets. No longer was it possible to use plug-in coils for band-switching, because you lost the shielding integrity, and effective shielding became almost impossible to sustain. It was important that the generated RF be contained so that the output could be fed through a low-pass filter to effectively kill any TVI-causing harmonics. It was much simpler to design the amplifier tank circuits for bandswitching using a pi network to work into the low-impedance filter designs. This meant coax feed lines. At first 70 ohm line was used and then 50 ohm line, simply because 50 ohm line was more readily available.

The Pi Network

The pi-network circuit lent itself to simple band-switching, plus it had a great deal of flexibility in dealing with wide ranges of reactance and impedance mismatches. Probably the best of those early commercial rigs was the Johnson Viking Ranger, which had the ability to match some really crazy loads. Gradually, however, the commercial manufacturers started to eliminate tank-circuit flexibility by designing transmitters with a fixed output impedance. This required that the amateur have an antenna load that was very close to 50 ohms impedance.

Practically every amateur wanted an antenna system that would be exactly 50 ohms impedance on every band and every frequency. In fact, I believe that manufacturers of transmitters really believed such a thing would come to pass, wherein an all-band antenna would become available with an impedance of 50 ohms (1 to 1 SWR on all bands, all frequencies). While there are some antenna systems that will give a very close match over a wide range of frequencies, there are none that match all bands. They just don't exist in the amateur marketplace.

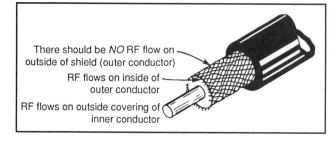

There should be *NO* RF flow on outside of shield (outer conductor)

RF flows on inside of outer conductor

RF flows on outside covering of inner conductor

Figure 2-2. A cross section of typical coax. The RF flows on the outside of the inner conductor, but more importantly on the inside of the outer conductor. There should be no RF flow on the outside of the line.

These are the latest of the new coaxial lines. They are covered in Table 2-1.

Keep in mind that the feed impedance of an antenna will change with height above earth, frequency, and so on. Adding to this is the fact that there are so many different low frequencies, it becomes an impossible task to come up with a 50 ohm impedance, all-band, all-frequency antenna. Keep that point in mind, because if you desire a perfect 1 to 1, 50 ohm load on all frequencies, then a Transmatch is required no matter what you read or hear. The Transmatch will be covered in detail in chapter 4. First let's take a closer look at transmission lines.

Transmission-Line Characteristics

All transmission lines are rated by their impedance. A transmission line's "characteristic" impedance is determined by the size of the conductors used in the line, the material composition used to hold the lines in place, and the spacing of the conductors. The most popular type of transmission line now used by amateurs is 50 ohm impedance coaxial line, certainly influenced by the history just discussed. Figure 2-2 shows the cross section of a coaxial line. The outside cover of insulating material can be made to withstand moisture or to permit actual burial underground for years without contamination.

One point that I will mention several times in this book is how RF flows in a coaxial line. RF from the transmitter flows on the *outside* of the inner conductor and on the *inside* of the outer shield. In the ideal situation no RF should flow on the outside of the outer shield. If it does, the outer shield will radiate because of a condition known as *parallel standing waves*—much more on that in a moment.

Getting back to SWR, assume we use this feed line to feed an antenna that has a 50 ohm impedance. Because the impedance of the transmission line matches the impedance of the antenna, the SWR will be a ratio of 1 to 1. This also means that the ratio of the maximum to minimum voltage at any point on the line is also 1 to 1. Keep this fact in mind: The impedance of the transmission line never changes; it is always 50 ohms anywhere along its total length. There is such a thing as a manufacturer's tolerance applied to impedance, and this is generally on the order of plus or minus 5 percent, but I will ignore tolerance in this discussion. Keep in mind that the line is made up of metal conductors, so it does have ohmic losses over its length (also the dielectric insulating material has losses which add to the overall losses). One last point to keep in mind is that these ohmic losses on the line *increase* as the operating frequency increases.

The reader desiring to choose the best coax for a job will find that since there are many types available, some basic information is required to do the job correctly. For example, most of us are familiar with RG-58U coax, but what the average amateur does not know is that depending on the manufacturer of the line, RG-58 can vary greatly as to loss and power-handling characteristics. RG-58, RG-8X, RG-8 are all lines that can vary significantly in quality. I am familiar with Times Wire and Cable, Nemal Electronics, Belden, Antennas West, The Wireman, and Transel Technologies, all of which handle transmission lines and antenna wire.

Table 2-1 was taken from a Times Wire and Cable brochure. If you are an amateur who has been around for a while, the loss figures per 100 feet on all the

CABLE DESIGNATIONS

Frequency	LMR200 CQ1022/RG58 0.195" dia. Atten./Max. Pwr	LMR240 CQ1004/RG8X 0.240" dia. Atten./Max. Pwr	LMR400 CQ1001/RG8 0.405" dia. Atten./Max. Pwr	LMR500 CQ1005/New 0.500" dia. Atten./Max. Pwr
30 MHz	1.8dB/670W	1.3dB/980W	0.7dB/2100W	0.54dB/2800W
50 MHz	2.3dB/520W	1.7dB/750W	0.9dB/1700W	0.7dB/2200W
150 MHz	4.0dB/300W	3.0dB/420W	1.5dB/1000W	1.2dB/1200W
220 MHz	4.8dB/250W	3.7dB/340W	1.8dB/830W	1.49dB/1000W
450 MHz	6.9dB/170W	5.3dB/240W	2.7dB/550W	2.17dB/700W
900 MHz	9.9dB/120W	7.6dB/170W	3.9dB/380W	3.1dB/490W

Table 2-1. This table of popular coaxial line types, specifically for amateur use, was taken from the Times Wire and Cable catalog (both The Wireman and Nemal Electronics carry these lines). Under cable designations, the first number is the Times Wire number followed by The Wireman's number, followed by the similar RG designation (i.e., LMR200/CQ1022/RG58). The attenuation figures given in this table are for 100 feet of line. These cables are rated for a 20-year lifetime and most can be buried. Your choice of cable depends on factors such as your installation, the frequencies to be used, and so on.

lines may surprise you. For example, RG-58 type line for years had losses on the order of 6 dB per 100 feet. Note that the newer RG-58 has a loss of only 4 dB, a considerable improvement indeed. Also note that losses in decibels per 100 feet of line are provided for various frequencies. In addition, maximum power ratings are shown.

Your installation depends a great deal on your goals. In order to be competitive you must try to keep losses as low as possible, and losses in terms of decibels can add up. The really outstanding stations always try to get every "edge" they can. If you are a ragchewer on 80 meters, then almost any transmission line is suitable. But if you are going for weak DX or a top contest score, then even a fraction of a decibel counts. And that means quality in choosing antenna components.

Always keep in mind that three decibels means a ratio of 2:1, or in other words, 3 dB loss means a loss

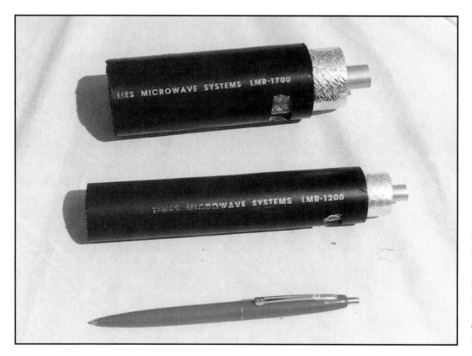

These are the LMR 1700 and LMR 1200 cables (manufacturered for The Wireman) for installations where very low-loss lines are required for VHF and UHF. The LMR 1700 has a diameter of 1.67 inches, and the LMR 1200 a diameter of 1.2 inches. The loss per 100 feet for the LM 1700 is very low at VHF and UHF. For 150 MHz, loss per 100 feet is only 0.347 dB, at 220 MHz only 0.427 dB, and at 450 MHz only 0.632 dB. The LMR 1200 checks out at 0.481 dB at 150 MHz and 0.864 dB at 450 MHz.

of half your power. The other handy number is 10 dB, which is a ratio of 10:1.

If I have any last thought, it is of the wonder of wire antennas. I grew up with them, and I normally used open-wire line to feed them. Coax definitely has a prominent place in our amateur lives, but we are always faced with SWR problems. Remember, if you learn nothing else from this book, an SWR of 30, 40, or even 80 to 1 means nothing with open-wire line. It is simply a matter of learning to use a Transmatch. With coax, try to keep the SWR below 2 or 3 to 1.

The most popular type of line used by amateurs may be coax, but as I have already pointed out, another type of line that is just as important is known as "open wire" feed line. The impedance of open-wire feed line is normally much higher than that of coax. Additional information on this feed line will be provided in chapter 3.

The Importance of Low SWR

If coax is going to be used with modern transmitting equipment, the SWR must be kept below a ratio of 2 to 1. The reason for this is that modern transmitting equipment includes built-in protective devices. These devices are engineered to keep the final amplifier transistor stages from overheating and blowing when a severe SWR mismatch is present. The limit of mismatch tolerated is usually an SWR of 2 to 1. In practice, what happens is the transmitter simply shuts down or reduces power when the mismatch it "sees" approaches or exceeds 2 to 1. It can quickly be seen that the coax feed line and the antenna must be matched if no other coupling device is used at the transmitter.

Not only should coax be matched for the reason just mentioned, but under some circumstances coax cannot tolerate high standing wave ratios. By high, we are thinking in terms of much over a ratio of say 5 to 1. When high power, 1000 watts or more, is used, coax must be kept at a low SWR. There is a danger of the line heating up and actually melting, or a danger of voltage arcing over, thereby puncturing the line at points of high voltage.

Because of these problems, antenna manufacturers have spent considerable time and effort to create antennas and matching methods which have a 50 ohm impedance to match the popular 50 ohm impedance coax, so that the desired 1 to 1 SWR ratio can be achieved. In fact, if one real problem exists in commercial antenna design it is that of sacrificing antenna gain and performance to achieve the 1 to 1 match. We will discuss this in detail when we go over multiband coax-fed antennas.

SWR and Bandwidth

Another important point in discussing the SWR of antennas is the bandwidth, or frequency range, over which the antenna will retain a 2 to 1 SWR ratio. A really simple fact of amateur life is understanding SWR changes as related to frequency changes. For example, assume we wish to make a coax-fed dipole for the 80 meter band—3500 to 4000 kHz. Suppose we make the dipole resonant at 3750 kHz, the exact center of the band, and install this antenna at a height above ground which will give the antenna an impedance of 50 ohms. This would provide a perfect 50 ohm match at 3750 kHz.

Unfortunately, 80 meters is a very large band in the sense of frequency ratio. If we look at the frequency ratio of the band, we find that it is 500 kHz wide, or a ratio of about 8 to 1 (4000 kHz divided by 500 kHz). Believe it or not, if we change frequency 100 kHz on 80 meters, we are introducing much more change than if we move 100 kHz on 20 meters. Again, the band on 80 meters is 500 kHz wide, 3500 to 4000 kHz. On 20 meters the band is 14,000 to 14,350 kHz, or 350 kHz wide. However, the actual frequency change *ratio* from one end of the 20 meter band to the other is a very small ratio—much, much less than that of 80 meters. This seems to be complicated, but it is rather simple to understand with some study.

If we actually measure this dipole antenna on the 80 meter band, we will find the SWR to be on the order of 10 to 1 at either end! This figure can vary because of other factors, but it won't be far off. Now on 20 meters, if we cut a dipole for 14,175 kHz, band center, the SWR is probably going to be no higher than 1.5 to 1 across the entire band.

What this means is that on 20 meters, for example, we go from 14,000 to 14,350 kHz, which is a much smaller ratio than on 80 meters. This brings us to a point which will bear repeating.

Amateur radio is a competitive as well as a pleasant hobby, make no mistake about that. If you and your buddy Joe get on the air and you both work the same station and he gets S9 and you get S8, you will be unhappy! If you are chasing DX and someone gets there first, you again will be unhappy! I have always worked on the assumption that wherever possible, get the very best signal out. If one coax is one-half decibel better than another, I'll take the one-half decibel. You'll be surprised how quickly a half decibel here and a half decibel there add up. Always look for an edge and never turn down a "little" edge! Those little edges add up.

RF Flow in Coax

Up to this point we have discussed coax line and the fact that the SWR must be kept reasonably low. Coaxial line, because of its shielding, keeps all of the RF inside the line—or at least it is supposed to. Refer again to figure 2-2. The RF from the transmitter flows on the *outside* of the inner conductor and on the *inside* of the outer conductor. If RF flows on the outside of the outer conductor, then a condition called *parallel standing waves exists,* and this will cause the feed line to radiate. This can be either a bad or a good condition, which will be explained later in chapter 6.

Keep one important point in mind when thinking of feed lines: A feed line is designed to carry the RF from the rig to the antenna without radiating. If the feed line radiates, it is not a feed line. It is an antenna! In fact, a good point to keep in mind is anything that radiates can be classified as an antenna.

SWR and Power Indicators

One item needed is an indicator which can show us when the antenna system is resonant. That indicator is usually an SWR indicator. There are many kinds of SWR indicators, ranging in cost from inexpensive to very costly. In any case, for our purposes usually the most inexpensive is just as good as the more expensive types.

It is possible to build an SWR power bridge, but the commercial units have become inexpensive enough so that I would not hesitate to recommend buying one. There are many varieties available, and they vary considerably in price.

One unit I have been using is the MFJ Model 249 HF/VHF SWR analyzer. This is a combination frequency meter and SWR indicator covering 1.8 through 170 MHz. I did a product review of this instrument in *CQ* magazine, so I had the opportunity to thoroughly evaluate its capability. Operation is very easy. The antenna feed line is simply connected to the unit and the frequency meter is tuned. A dip in the SWR indicator meter reading indicates resonance. An excellent article in the November 1993 issue of *QST* described a small probe that converted the MFJ-249

This is the MFJ-249 SWR/dipper/frequency counter described in the text.

into a type of "dip" meter. This modified unit can be used to check guy lines, outside of coax, etc., for resonances—an extremely useful function.

In this same vein I must also mention another useful device, the Bird Wattmeter. The Bird unit has more or less become the standard power and SWR indicator in the amateur field. It is an extremely accurate device for measuring power, and in fact is used by the FCC Field Division for many measuring checks.

In addition, there are many useful test units that can be home made which are just as effective as commercial devices. Just a little more history on this subject before I show you how to build an excellent home-made power and SWR bridge.

The Monimatch

Coaxial line become popular after World War II. Amateurs started using the line because it was cheap, it could be found on the surplus market, and it was a very handy shielded feed line. As I explain elsewhere in this book, early amateurs were not concerned with SWR ratios simply because it was almost impossible to measure the waves.

In 1957 I created a circuit called the Monimatch.

Figure 2-3. This is the circuit of the GM4ZNX bridge, which does not use nulling capacitors. All capacitors are disk ceramic and are in μF. Resistance is in ohms. R5 and R6 are PC-mounted potentiometers. See the text for D1 and D2 information. RFC1 and RFC2 are miniature molded RF chokes. T1 and T2 have 20 turns of No. 26 enameled wire on each transformer for power levels up to 1500 watts. Use 12 turns of No. 24 enameled wire on each transformer for QRP operation. T1 and T2 are wound on Amidon Associates FT-50-61 ferrite toroids for operation from 3.5 to 30 MHz. Use FT-50-43 cores for 1.8 through 30 MHz.

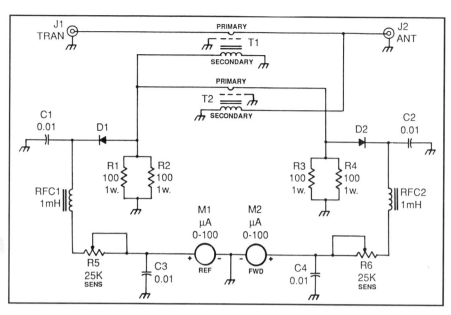

The Monimatch, a simple reflectometer, was an easy to build device which could actually measure the SWR voltage ratios present *inside* coaxial transmission lines. Because of its simplicity and low cost, the Monimatch became extremely popular and made everyone more aware of SWR and its problems. The Monimatch had two problems, though. First it was frequency sensitive, and because of this it was not great for power measurements. It could be calibrated accurately for a single band, but as one went higher in frequency, it needed to be calibrated again and again if power was to be measured accurately.

The Breune Bridge

An engineer with Collins radio, Walter Breune, came up with what eventually became known as the Breune Bridge. This unit could be calibrated accurately for power measurements and was not frequency sensitive. The bridge soon became extremely popular. It did have a drawback, however, in that it was rather difficult to obtain an *exact* balance between reflected power and forward power. In addition, everyone started referring to *reflected power,* a term still misunderstood by many amateurs. That, however, is another story.

The DeMaw Bridge With Construction Details

Many versions of the Breune Bridge have been built improving on the first design. A large number of these—including the very latest version of a unit that gets rid of all the "accuracy" problems—were

designed by Doug DeMaw, W1FB. The above unit was first concocted by D. Stockton, GM4ZNX, and appeared in the British QRP Journal, *SPRAT*. Doug DeMaw described this unit in the June 1994 issue of *CQ* magazine. Following is a modified quote from Doug's article.

The circuit for this relatively fool-proof bridge is presented in figure 2-3. I built several versions of Stockton's circuit and was delighted with the inherent balance. No balancing capacitors are required for the design. The primary criterion is that the layout be symmetrical, and that the leads are kept as short as possible.

The circuit in figure 2-3 may be tailored easily for QRP or QRO use. The sensitivity is determined by the turns ratio

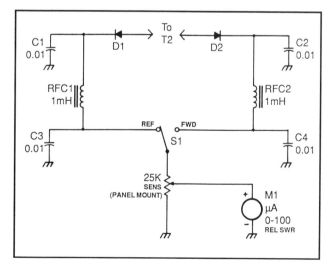

Figure 2-4. Example of a method for using one meter, a FWD/REF switch, and a sensitivity control for relative SWR readings.

The board, which is only 3 inches by 3 inches, does not take up much room. The board should be mounted in a metal enclosure for shielding. If the unit is mounted in a Transmatch, be sure to use shielded leads from the board to the meter.

resistors have become as rare as teeth in an earthworm these days. Modern, so-called carbon resistors are now known as carbon filament resistors. The resistive element is spiral wound over an insulating form and does exhibit a small amount of unwanted inductive reactance. However, I have observed no bad effects when using them in critical circuits up to 30 MHz. They can present problems at VHF and higher. Carbon film resistors are suitable in this circuit. I paralleled two 100 ohm units at each load point in order to minimize the internal inductance of the resistors while providing the internal 50 ohm loads for the circuit.

The short lengths of 50 ohm RG-58A coax that pass through the toroidal transformers, T1 and T2, have the shield braid grounded at one end, as indicated in the diagram. This provides Faraday shielding for best bridge performance.

Separate meters are used to monitor both forward and reflected current simultaneously. If desired, you may prefer a single switched meter. The circuit for a single meter is shown in figure 2-4. R5 and R6 are adjusted to provide full-scale meter deflection in the FWD mode for a specified power level. The bridge should be terminated in a 50 ohm dummy load for this calibration procedure. If desired, the meter face can be marked for the various power levels by measuring the RMS voltage across the dummy load with a VTVM and an RF probe (P watts = E^2/R) while varying the transmitter output power. Or, the scale can be set by using a calibrated RF wattmeter for comparisons. M1 and M2 are 100 uA DC meters. One can use meters up to 500 uA,

of the two transformers, T1 and T2. As to the diodes, I have had good luck with matched 1N914s at 100 watts or greater. For QRP I use 1N34A diodes, which work nicely.

R1 through R4 in figure 2-3 should be 1 watt carbon composition resistors. Unfortunately, these noninductive

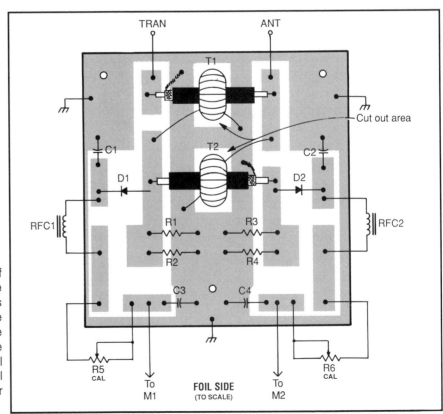

Figure 2-5. This is a layout to scale of the bridge PC board showing where the components are located. The parts are installed on the foil side of the board. PC boards for this project are available from FAR Circuits. It is simple to make your own board with the usual etching procedure. Use a motor tool and routing bit to form the copper islands.

except that they will not work for very low QRP work.

A PC-board pattern for the GM4ZNX bridge is shown in figure 2-5. If you are experienced at laying out PC boards, you may wish to make the circuit more compact. It should be noted that the ground conductors on the board should have good electrical contact with the metal chassis or box and with the module to which it is attached. It is suggested that short metal standoffs be used at the three circuit board mounting holes to ensure good ground connections.

Testing of the bridge is accomplished by attaching a 50 ohm noninductive load to J2. RF power is applied at J1. Adjust the transmitter output power to obtain a full-scale reading at M1 with the bridge set for the FWD mode. Now switch to read REF power. M1 should have a reading of zero. Reverse the connections at J1 and J2, select the REF mode, and apply RF power. M1 should have a full-scale reading and should indicate zero in the FWD position of the switch. These readings indicate correct bridge balance. If these results cannot be obtained, you will need to make certain that the wiring is correct, that D1 and D2 are closely matched, and that R1 through R4 each have a resistance of 100 ohms. Also be sure that T1 and T2 are connected exactly as shown in figure 2-3.

You can match your diodes by checking the forward and reverse resistance with an ohmmeter. The forward resistance (low reading) is the most important one to consider. Generally, it will be on the order of 5 to 10 ohms for 1N914 diodes. The back resistance will be on the order of over 100K ohms. A digital ohmmeter will provide the best results for this test.

The above description by Doug DeMaw is an excellent guide for construction of this bridge. SWR bridges are one of the most useful devices in the shack. They show when a Transmatch or antenna tuner is correctly adjusted. They show if power is actually being put out from the rig. They show the standing wave ratio on a line. They show when something is not normal with the antenna. In other words, they are indispensable.

Chapter 3

Decibels

In order to discuss antenna power gain, it is important to get some basic facts into the record. For example, to achieve an antenna gain of 10 dB, amateurs sometimes think that the power must be increased by a factor of 10. In other words, a 100 watt signal becomes a 1000 watt signal. This is both true and false, if you can follow crazy logic. What happens is *some antennas can concentrate their patterns in given directions much better than others.*

Let's assume that an antenna takes a 100 watt signal and radiates 100 watts in all directions. Then assume that we modify the antenna so that the 100 watts of power radiated by it is concentrated in one specific direction. Figure 3-1 is an illustration of the horizontal pattern of a dipole at A and a beam at B. If the maximum field strength in the main direction of the beam pattern has a gain of 10 decibels in relation to the dipole pattern at A, then a significant signal strength increase exists in the desired direction. We are not actually increasing the transmitter power, but we are shaping the pattern to direct power in a specific direction, which effectively acts as increased power in that direction.

Antenna Patterns

All antennas have some kind of pattern. The theoretical isotropic antenna—and it is only theoretical—is an antenna that radiates equally well (or equally poorly) in all directions. The 80 meter dipole I mentioned earlier, if it were not affected by ground or if it were in free space, has a figure-8 pattern. Visualize, if you will, a doughnut. If the dipole were placed through the center hole, the doughnut shape around the dipole would simulate the antenna pattern. However, and this is what is important, that shape would only exist for a resonant half-wave dipole.

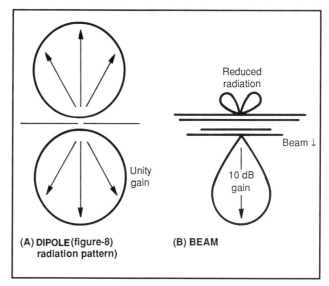

Figure 3-1. At (A) is the pattern of a dipole in the horizontal plane. Note that the signal is maximum, broadside to the plane of the dipole, with little or no radiation off the dipole's ends. Section (B) illustrates how a beam modifies the pattern by putting a strong lobe of radiation in one direction, with very little in the rear direction. Ten decibels of gain can be assumed when compared to the dipole.

If we use the 80 meter dipole on 20 meters, the pattern will now resemble two butterflies, wings spread, placed head to head (see figure 3-2)—or in other words, four lobes of radiation. We have taken the figure-8 doughnut pattern and shaped it into a pattern that has four lobes. These lobes concentrate more energy in certain directions than the figure-8. Depending on several factors, these lobes could have three decibels of gain compared to the 80 meter lobes.

Decibels and Antenna Gain

Now let's discuss decibels. Antennas are rated in per-

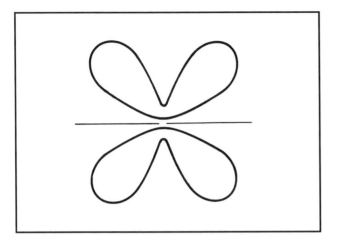

Figure 3-2. Typical antenna radiation pattern of an 80 meter dipole operated on 20 meters. It is described in the text as two butterflies, wings spread, placed head to head. Note the four lobes of radiation.

Decibels (dB)	Power Change	Decibels (dB)	Power Change
1	1.25	10	10.0
2	1.58	11	12.6
3	2.0	12	15.8
4	2.5	13	20.0
5	3.15	14	25.1
6	4.0	15	31.6
7	5.0	20	100
8	6.3	30	1,000
9	7.9	40	10,000

Table 3-1. Change in power as measured in decibels.

formance by gain, and gain is measured in decibels, noted as dB. Antenna gain is normally compared to a half-wavelength dipole in free space. Losses in feed lines are also given in decibels, so the reader needs an understanding of dB. For convenience, we have provided a chart showing decibel gain versus power (table 3-1). In audio, for example, the slightest detectable difference in audio level is defined as one decibel. In radio there are a couple of very simple decibel ratings that have meaning to antenna people. A gain of 3 dB in a given direction is the same as doubling the power in that direction. A gain of 10 dB is the same as increasing the power 10 times. Going from 100 watts output to 1000 watts output is a gain of 10 dB. These days we normally think that S meters (signal strength meters) on transceivers have a difference of 6 dB per S unit. The actual difference can vary with different receivers, and even from band to band. However, the 6 dB per S unit is a good rule of thumb. *Important:* We are assuming here that a receiver's S meter is very accurately calibrated.

A very interesting antenna discussion is that of gain and increasing gain, particularly with beams. It is generally accepted that a three-element Yagi monoband beam has very close to 7 dB of gain compared to a half-wave dipole. Innumerable measurements confirm a 7 to 7.4 dB gain for a three-element beam. To increase that gain by 3 dB (double the effective radiated power), we simply *double the size of the beam.* In other words, for the sake of discussion let's state that we have a three-element beam on a 14 foot boom with 7 dB gain. In order to obtain a 10 dB gain antenna, we need six elements on a 28 foot boom! To carry it to the absurd, to get 13 dB (an additional 3 dB) we must double again to a 56 foot boom with 12 elements. We could double in another fashion based on what VHF and UHF enthusiasts do, by building two 7 dB beams and stacking them a short distance apart. No, we don't get twice 7, or 14 dB. We get 10 dB (remember the increase of 3 dB).

The point here is that when you see advertising figures of 11 dBd (decibels compared to a dipole), ask yourself if the antenna has the necessary number of elements and the boom length to justify the claim. Remember those two decibel figures: 3 dB is double the power (and double the size in a beam) and 10 dB is 10 times the power.

Transmission Lines and Decibels

Despite the current interest in coax line, in the early days of radio, amateurs made their own transmission lines from two lengths of wire, usually No. 12 solid-copper enameled wire. These two wires were run in parallel and separated up to as much as 6 inches by insulated spreaders. Two inches of separation was, and is, a popular dimension. Many amateurs still use such lines. In recent years a new type of open-wire line has become very popular. Both the old-fashioned and new are shown in figure 3-3. The newer line is insulated with a poly covering, and the wires are spaced approximately 5/8 inch apart. The insulation is opened every few inches with air separating the wires. This gives the line an appearance of a ladder, and it is commonly called *ladder line.* Such lines are extremely low loss and can tolerate very high SWRs (and therefore high power). In my station I use both coax lines and open-wire lines. I use coax to feed my beams and open-wire line for a multiband dipole.

What are the advantages and disadvantages of open-wire line? Probably the primary advantage of open-wire line is the fact that it is practically "lossless." In both coaxial lines and open-wire lines as the SWR increases so do the losses, but the losses can be appreciable under certain conditions *only* when using coax.

One of the most startling examples of the above is in feeding a dipole that has an impedance that matches the coax, such as a 50 ohm dipole fed with 50 ohm line. Let's use an 80 meter dipole as an example and assume an impedance of 50 ohms. We would have a 1 to 1 SWR feeding the antenna with 50 ohm coax. But let's also assume that we wish to use this same antenna on the next higher band—40 meters. The 80 meter dipole makes an excellent full-wave antenna on 40 meters. In fact, as a full-wave antenna it will exhibit some gain when compared to the half-wavelength 40 meter dipole.

The typical figure-8 pattern will change when we use this antenna on 40, but more important, a problem exists in the feed-point impedance. For a half-wavelength dipole the impedance will be 50 ohms, but for a full-wavelength dipole the impedance changes radically. The full-wavelength dipole will have an impedance in the vicinity of 4000 ohms! It does not take much mathematics to realize that the 50 ohm impedance line feeding a full-wavelength dipole will have an SWR of 80 to 1! The 50 divided into the 4000 ohm impedance gives us the 80 to 1 SWR.

The normal loss for 100 feet of coax at 3.5 MHz (80 meter band) is only 0.3 dB, an insignificant figure. But because losses in a line increase as the SWR increases, the losses in this case will go over 10 dB. (Ten decibels just happens to work out to a ratio of 10 to 1, so quite simply, putting 100 watts into the line at the rig would result in only one tenth of the power—10 watts—reaching the antenna.) Also, because SWR is a ratio of voltage or current at any point in the line, an SWR of 80 to 1 would mean that extremely high voltages exist at many points in the feed line. This certainly can blow out a coax line if high power is used. In addition, the high SWR also means high currents are present, which could cause the line to heat or melt.

On the other hand, open-wire line is so lossless that an 80 to 1 mismatch is no real problem. In addition, the impedance of the open-wire line is on the order of 450 ohms, so the mismatch would only be about 10 to 1 anyway. This is merely an example to show what can happen.

(I cannot help but mention here that many years ago I gave an antenna lecture and pointed out this 80/40 meter coax problem. One fellow in the audience raised

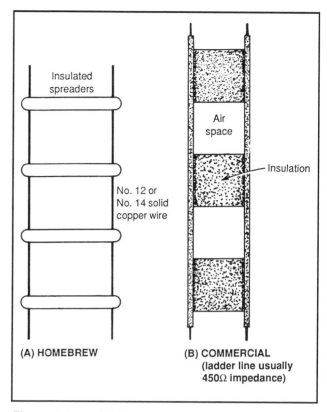

Figure 3-3. At (A) is a typical piece of open-wire line. Insulated spreaders are used to hold the two wires parallel with separation of 2, 4, or 6 inches. The impedance of the line is determined by the wire size and the spacing of the lines. At (B) is the very popular open-wire line. It is not really "open" wire because it does have insulation around it. The better grades of this line will easily handle the legal power limits.

his hand and said, "I use just such an antenna and feed it with coax, but I work out very well on 40. What's your answer to that?" I paused for a second or two and said, "Well, then you use McCoy's first rule on antennas: If the darn thing works, don't change it!")

A more realistic example, and one that applies to many newcomers, is using RG-58/U to feed beams or antennas on 10 meters. RG-58 is a lower cost 50 ohm coax, and it is rated at 2.2 dB loss per 100 feet on 10 meters. With a mismatch of only 2 to 1, the losses will go to about 3 dB, and 3 dB, as we pointed out earlier, will cut the power in half. This is not to say you should not use coax to feed a beam. In fact, it is the logical thing to do. However, what is false economy is to use a cheap, high-loss, coaxial transmission line. There is absolutely no sense in throwing away half the power, when for a relatively small cost the problem can be avoided.

We stated that open-wire line is low loss, but how

low is low? The newer of the "covered" open-wire lines will show a loss of only 0.30 dB per 100 feet at 10 meters. Back on 80 or 40 meters, the losses of this type of line are, for all practical purposes, meaningless. The loss at 80 meters for 100 feet is only 0.1 dB. For all practical purposes again, it is a lossless line. Our earlier 80 to 1 SWR wouldn't be worth considering with this type of line.

Table 2-1 in chapter 2 lists coax types with the losses for all of the more popular lines. Note that some of the coaxial lines, which are fairly new at this date, are extremely low loss even at VHF and UHF. Special consideration must be given to feed-line losses at VHF and UHF. The ordinary RG-8/U coaxial line has a loss of nearly 5 dB per 100 feet at 450 MHz. Just to translate that number into real figures, let's suppose we are running an FM base station to an antenna 100 feet away and 50 watts input to the transmission line. We will lose about 35 watts on the way to the antenna; approximately 70 percent of the power would be lost heating up the line! Of course, the answer is to choose as low-loss a feed line as possible.

While we have pointed out that open-wire line is very low loss, it is not a good choice for VHF or UHF. The problem is that whenever the conductors are relatively widely spaced, as they are with open-wire line, there is always the possibility that the line itself can radiate—an undesired condition.

For frequencies below 144 MHz, however, open-wire feeders offer many possibilities for multiband use where high standing wave ratios can occur. This leads us to the methods for using open-wire line, coupling methods to the transmitter, and how to get your feeders into the shack.

Chapter 4

Transmatches

The logical answer to many of the problems connected with SWR and power loss discussed in chapter 3 is the use of a Transmatch. Exactly what is a Transmatch? The name was coined by George Grammer and myself in my article "The 50 Ohmer Transmatch." We created the name to avoid the use of the name *antenna coupler* or *tuner*, which is not appropriate because you do not tune an antenna. In practice, the antenna system—a combination of the antenna and the feed line—is an unknown load which is tuned to match the transmitter output impedance.

The Unknown Load

One of the most difficult technical points for a newcomer to understand is the load that exists at the end of the feed line connected to the transmitter. Many amateurs think that when a 50 ohm impedance line is

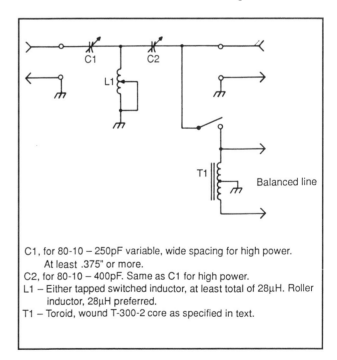

C1, for 80-10 – 250pF variable, wide spacing for high power. At least .375" or more.
C2, for 80-10 – 400pF. Same as C1 for high power.
L1 – Either tapped switched inductor, at least total of 28μH. Roller inductor, 28μH preferred.
T1 – Toroid, wound T-300-2 core as specified in text.

Figure 4-1. Three versions of the Transmatch. At (A) is the most popular of the Transmatch types. It includes two variable capacitors and a variable inductor. Shown at (B) is the Ultimate Transmatch and at (C) the SPC. Any one of these three, using proper values for the variables, will match any load an amateur is likely to encounter. The circuit at (A) is preferred because of its simplicity.

Figure 4-2. This is the circuit of the Transmatch for multiband use shown in figure 4-1(A). L1 typically would be 25 to 30 μH, and the band taps would be determined with the multiband antenna connected to the output. Before making permanent taps on the coil, use a clip lead to short out turns, adjusting C1 and C2 for a match. Use the tuning techniques outlined in the text. T1 is the transformer described in figure 4-5 and the photos.

attached to the rig, the load is going to be 50 ohms. About the only time such a condition exists is when the antenna impedance is exactly 50 ohms and is non-reactive. In truth, the load can vary from as low as a fraction of an ohm to several thousand ohms! Therefore, the problem becomes one of converting this unknown load to a pure 50 ohm load.

Essentially, a Transmatch can be called an adjustable RF transformer and reactance "tuner-outer." It takes the unknown load at the transmitter end of the coaxial line and transforms that load to 50 ohms—the normal output impedance of the transmitter. Note that I said "unknown load." The value of the load can be measured with the proper equipment, but that isn't necessary. What is necessary is that we cancel out any reactance present and step up or step down to achieve the match needed. The circuit designs of typical Transmatches are shown in figure 4-1. The most common currently in use is 4-1(A), which is further detailed in figure 4-2.

The above tends to emphasize the inherent problems in using coax with a Transmatch and the handling of high SWRs. But what about open-wire line and a Transmatch? As pointed out, open wire is an extremely low-loss line which can tolerate almost any SWR, no matter how high the ratio. I have used open-wire line consistently with an SWR of at least 80 to 1! Always keep in mind that if a line is essentially loss-less, then the power must go to the antenna to be radiated. Feed lines do not radiate simply because of a high or low SWR. There are some instances of feed-line radiation that will be discussed in detail later, but regardless of what you may hear on the air, SWR does *not* make a feed line radiate.

That being the case, the use of open-wire line and a Transmatch opens an efficient world of wire antennas without using complicated matching devices at the antenna which consume power. This book's aim is to show good, practical technical concepts which can provide a little gain here and a little more there regardless of the antenna location. Remember that every incremental improvement of an antenna's performance is important whether applied to an indoor antenna in an enclosed building or (my favorite location) to a salt marsh on a 6000 foot plateau!

Transceiver Manufacturing Limits

These days many transceivers have built-in Transmatches or tuners. Most of these have very limited matching ranges, usually on the order of 25 to

150 ohms, and most of them do not have provisions for balanced loads (open-wire line). The question very frequently asked is "If my transceiver has a built-in Transmatch, do I still need a regular or separate Transmatch?" Obviously, if there is a load that is outside the range of the built-in unit, then a wider range circuit is required. This brings up the subject of what kind of tolerance in SWR is acceptable to modern transceivers.

Most of the modern transceivers have built-in protective circuits that will keep the final amplifier stages from being destroyed if the mismatch becomes too high. What is too high? From many tests it was determined that if the load to the transceiver goes higher than 2 to 1, the transmitter will automatically shut down. In fact, in some tests when the SWR started to go over 1.5 to 1, the output began to drop. From a realistic power-loss standpoint, an SWR below 2 to 1 is almost unmeasurable, so we certainly can accept such a load.

When Is A Transmatch Needed?

A good question frequently asked is "When is a Transmatch needed?" Naturally, if the transmitter is in a shut-down condition because of a high SWR, we will need to either match the antenna to the line or use a Transmatch. Matching the antenna to the line can be an impossible situation in many cases, particularly when we QSY, and so forth. Therefore, in this case a Transmatch is the answer.

Also, don't be misled by claims from makers of multiband trap dipoles that state the antenna will stay below 2 to 1 on all amateur HF bands, 160 or 80 through 10. This cannot happen unless very high-loss resistors are used. To reiterate, a Transmatch is not needed if the SWR stays below 2 to 1 (but frankly, I prefer to use one anyway, because my amplifier will always be working into a matched condition). On the other hand, I might add that multiband trap dipoles and other antennas such as the off-center-fed models will present a "reasonable" antenna system load for a coax-fed antenna to work with a Transmatch. (A Transmatch will be needed to use all frequencies.) In other words, we don't want to use coax as a "tuned" line if the SWR is really high, as it can be with some antenna impedances. With multiband coax-fed dipoles we can expect the SWR to be a reasonable figure. It should also be kept in mind that antenna efficiency will always suffer to some degree if traps are used.

Without using a Transmatch, what is an "ideal"

SWR? Based on the manufacturer limits covered earlier, I must say the ideal is 2 to 1 or less. Any loss in the transmission line from a match of 1 to 1 up to 2 to 1 is insignificant and is not considered important. I would say then that if the antenna system feed stays below 2 to 1 on our operating bands and frequencies, we may not need a Transmatch. Because of the transceiver 2 to 1 restriction, however, we do need a Transmatch if this figure is exceeded.

Another bonus when using a Transmatch is that it will offer a certain degree of selectivity to the station. For example, amateurs living close to broadcast stations or other high-power RF installations can get a lot of interference from cross-modulation generated by the overloading of the receiver by these high-power signals. In many instances a Transmatch will eliminate this problem.

Because of the Transmatch selectivity, there is a certain amount of harmonic rejection. However, I do not intend to get into a discussion of harmonic suppression and which of the common Transmatch circuits—the T, SPC, or Ultimate shown in figure 4-1—is best for harmonic reduction. Each has its advantages, and in any case, the argument about harmonics is academic simply because the FCC rules state that harmonics must be down at least 40 dB from the final stage of the transmitter for the 160 through 10 meter bands, and as far as I know, all commercially built rigs meet this standard.

Regardless of which of the above-mentioned circuits you use, all have the infinite beauty of matching any kind of antenna system. Think about that statement for a second. I said *any kind of antenna system!* This means random-length wires, rain gutters, guy lines on towers, or towers themselves. In fact, anything that is metal can be matched, to a 1 to 1 ratio, with the circuits mentioned above. I guess I should add that most of the commercial units using the above-mentioned circuits are usually for 80 through 10 meters (some also include 160 meters).

To satisfy the statement of matching "anything" requires having the three main components—input capacitor, inductor, and output capacitor—all variable. Many of the commercial circuits use a switched tapped inductor. While this is okay in most matching cases, there are some loads that are impossible to match perfectly. Usually, however, a better than 2 to 1 match for any system can be achieved.

Another argument that has ensued over the years pertains to the use of roller inductors in Transmatches. The original circuit I designed used a variable inductor

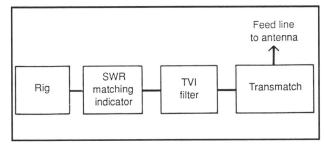

Figure 4-3. This is the normal setup for using a Transmatch and an SWR bridge. Sometimes the SWR indicator is built into either the transceiver or the Transmatch.

in the form of a coil that could be rotated, shorting out turns and changing the inductance as it was rotated. There are Transmatches that use tapped inductors whereby turns are shorted out by means of a switch to disconnect unused turns. In many instances an antenna system requires a very *precise* amount of inductance which cannot be achieved by the use of tapped turns. The *ideal* Transmatch uses such a completely variable inductor simply because it can be set precisely.

Some amateurs are concerned about a Transmatch introducing loss into the system. Some years back I made extensive tests and found that when using good RF connections on all components in a Transmatch, plus reasonable power-handling components, the average power lost was on the order of 3 to 5 percent, depending on the load being matched. On the other hand, this loss can be more than compensated for by using the Transmatch so that a rig is always working into its design load. Everything runs cooler, with better efficiency, and so on.

Both sides of this coin must be looked at, however. What is gained? Keep the following point in mind. A transmitter is designed to work into a 50 ohm impedance line. By using a Transmatch, we ensure that the rig is always working into the proper design load, producing the most efficiency! Any insertion losses are therefore quickly overcome.

The Ultimate Transmatch

Many years ago I designed a unit known as the Ultimate Transmatch. It was very similar to the circuit shown in figure 4-1(A) and is shown in figure 4-1(B). The problem I faced was how to get a basically unbalanced output to feed a balanced line

such as open-wire feeders. I realized first that the antenna system load, although the feeders were 450 ohm impedance, would be, or could be, some value

This is the heart of the ICOM AH-2, an automatic tuner. It is capable of handling 200 watts under varying loads. It is primarily designed to remotely tune a multiband whip such as a mobile or RV installation. I thoroughly tested and reviewed this unit and rate it very highly.

very far removed from 450 ohms. Baluns are designed to go from unbalanced to balanced loads, but in this case, although a balun might have worked, I was dealing with wide and varied impedances. If I used a balun, say a 4 to 1 ratio balun, I might have been able to convert that balanced load to an unbalanced coupling system. But again, I couldn't think in terms of baluns. I had to think strictly in terms of transforming an unknown load to 50 ohms. This presented crazy problems because of the extremely high SWRs I could encounter. High SWRs can cause high heating problems, as I soon discovered.

In any case, I started out with a 4 to 1 basic transformer that used a T-200 powdered iron core. This worked fine at power less than a few hundred watts with some loads, but with other loads it heated up. I ran more power as a test and promptly shattered the powdered iron core. To make a very long story short, after many days of work I finally ended up with three stacked, special insulation wrapped, T-200 cores wound with No. 14 Teflon covered wire. The system worked with all loads and was tested at 2000 watts. That particular transformer (call it a balun, if you like) has more than withstood the test of time. Construction details are presented later in this chapter.

Arguments erupted: I should have used 1 to 1 transformer ratios, not 4 to 1, or the cores saturated and generated spurious signals, etc. All of these statements have proven to be false over the years. Dr. Jerry Sevick, W2FMI, who is the authority on ferrite and powdered iron transformers, completely vindicated

my transformer in an article in the November 1993 issue of *CQ* magazine. My selection and design is also supported by the fact that most of the Transmatch (antenna coupler) manufacturers are using my circuit.

Installing A Transmatch

Before discussing the technique used to adjust a Transmatch, let's see where a Transmatch should be installed. (See figure 4-3 for the basic setup of a Transmatch installation.) If a low-pass filter is used, it should be installed between the rig and the Transmatch, and also between the SWR bridge and the Transmatch. In extreme-fringe areas of television special care must be exercised. The diodes used in bridges can generate harmonics, which the low-pass filter must "kill." The SWR bridge placement is not important in cable TV areas or in strong signal locations, just in fringe or weak-signal spots. Filters are designed to work into 50 ohms, and that point in the station setup is where a 50 ohm impedance will be.

There are many commercially available "automatic" Transmatches, and many modern transceivers have the option of providing a built-in automatic Transmatch. I cannot get too excited about most of these units simply because the matching range is very limited. Amateurs purchase such transceivers expecting them to match any antenna load and are disappointed when they don't. On the other hand, there are some automatic tuners that handle practically any load. The accompanying photos show two of the units that will handle just about any load an amateur will encounter. To be fair, there are other commercial units available. The questions that should be asked are simple enough: "What is the matching range of impedances covered and what is the power level they will handle?"

An SWR indicator bridge should be placed between the rig and the Transmatch as shown in the photos. This is important, because the bridge provides the visual indication of when the system is correctly tuned. Many commercial units have an SWR bridge built in, so note what I said about weak TV signal areas. Keep notes as you do your adjusting, because once you find the correct settings, you will want to make a record of them so you can quickly return to the proper settings whenever you change frequencies.

Transmatch Tuning

Now for the best technique to tune or adjust a Transmatch. First, and most important, a Transmatch must be adjusted using the minimum amount of power that provides indications. If the SWR indicator has different power-level settings, the lowest one should be used for adjustments. Assume we are using a multiband Transmatch, 80 through 10 meters. Let's review the procedure for 80 and 40 meters first.

Before applying power, the variable capacitors should be set to maximum capacitance—plates fully meshed. It is not unusual to find several different Transmatch settings which provide a perfect match. Always use the match that provides the most capacitance in the circuit. I could go into the whys, but that would be unnecessarily technical at this point. Just take my word for it: **Maximum capacitance is always best.**

Next we switch the bridge to read FORWARD, turn on the rig, and adjust the drive or gain control to put out enough power for a reading. Usually 10 to 20 watts with a 100 watt rig is sufficient. This level of power could be used all day without hurting anything. If you have a roller inductor for your inductance, start at minimum inductance and run the roller out slowly, increasing the inductance while observing the bridge meter. At some point you will get an indication of more power output.

Now we switch to the REFLECTED reading and adjust both capacitors, looking for a drop toward zero. The closer the reading gets to zero, the closer we are to a 50 ohm match. Next we carefully adjust the inductor and the capacitors, looking for a zero reflected reading on the meter. Once we reach zero, we have

Here is the Ten-Tec fully automatic tuner capable of matching virtually any load, covering both coax and open-wire or single feed lines. It covers 160 through 10 meters. Power-handling capability is the full legal amateur limit (1500 watts). It has 21 memories that can be preset, so band changing is fully automatic to a 1 to 1 matched condition in seconds.

matched the antenna system, an unknown load, to 50 ohms, the design factor of your transmitter.

We now bring up the power to the rated transmitter level. The reflected reading may have to be touched up with the Transmatch controls. (If you are operating on 80 meters, record your frequency and QSY to see how far you can move and still stay less than the magic 2 to 1. Usually, with an 80 meter half-wave dipole fed with open-wire line you will only be able to move about 100 kHz before an adjustment of the Transmatch is required.)

The procedure for 20 through 10 meters is similar, except that the capacitors will be set at one-half or one-quarter mesh for the proper setting. Also, only a few turns of inductance are needed on these bands. In fact, on 10 meters matching some loads is touchy, but it can be done if care is used. With switched, tapped inductor Transmatches different switch settings will have to be tried. With some loads 1 to 1 may not be able to be reached, but usually very close to a good match can be achieved. Personally, I accept a match of 1.5 to 1 or better.

By using a Transmatch we now have a "tuned system." It is important to know that neither the SWR on the transmission line nor the pattern of the antenna

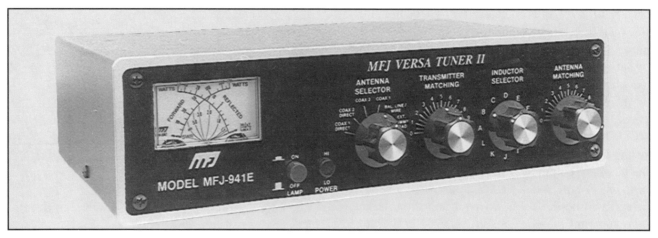

This is a 150 watt class commercial Transmatch made by MFJ.

has been changed. We have taken the antenna and feed line, an unknown load or unknown reactance, and converted that load to one which the transmitter and receiver see as a "pure" 50 ohm load. Admittedly, this is over-simplification, but it gets the job done. Also, and this is very important, any dipole of any length, fed with twin lead or open-wire line, is a true multi-band system—no traps, no baluns, just the feed line and the dipole.

As mentioned elsewhere in this book, the question I frequently am asked is "Do you always use a Transmatch?" The answer is yes, simply because of the modern transmitter and amplifier design. I *always* want to see a perfectly matched load even though I

know that the low SWR doesn't have much loss. Too many other variables can get into the act. With a Transmatch, however, I know I can always work into the design load.

At this point I would be remiss if I didn't state that open wire should not be used to feed a directional rotary beam—at least not without giving the problem some thought. Any very long runs of coax line can produce some bad losses which can negate the gain of a beam. However, in some cases amateurs find it necessary to install their beams at considerable distances from the shack, which would be the basis for the high coaxial losses mentioned. For runs of over 250 feet I would suggest going to open-wire line to get to the

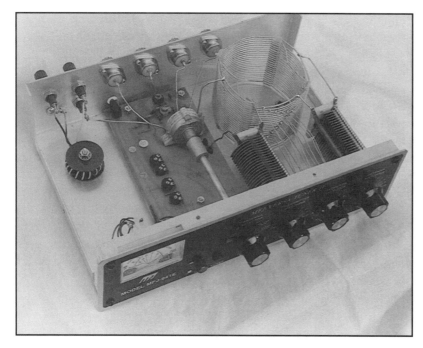

The interior view of the MFJ Transmatch. They use a tapped inductor in this version. If you have one and cannot get a perfect match, try using a clip lead to find the correct tap point for a perfect 1 to 1 match. Somewhere on the coil there is a tap that will get you to 1 to 1.

beam, then using one of Jerry Sevick, W2FMI's 9 to 1 balanced to unbalanced transformers (see Jerry's new book, *Building and Using Baluns and Ununs*, published by CQ Communications). This would get you from the 450 ohm impedance of the open wire to 50 ohm coax, permitting you to feed the beam directly with the coax.

I cannot stress too much how important Sevick's book is to amateurs interested in making baluns or transformers to get from one impedance to another. I could cover the construction of a 1 to 9 (50 to 450 ohm) transformer/balun in this book, but there is more benefit in studying Sevick's book.

In the case of a multiband dipole such as the McCoy dipole or the G5RV type (both discussed in detail later), we would definitely want to use open-wire line. With it extremely high mismatches can be tolerated without loss. This cannot be done with coax.

That leads us to the discussion of feeding a multiband dipole with open-wire line. When I say a "multiband" dipole, I mean a wire dipole opened at the center and fed with open-wire line (or it could be an end-

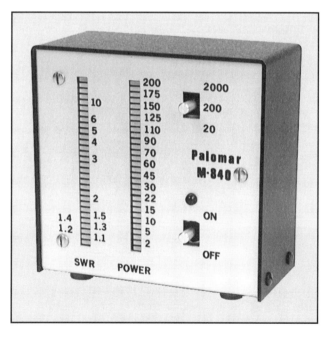

Certainly one of the finest tune-up indicators for a Transmatch or tuner is the Palomar Engineers Model M-840. This unit is an SWR/power bridge that shows the power in three switchable stages—20, 200, or 2000 watts. Bars of light indicate the levels of both power and SWR. I have found that tuning up a Transmatch is very simple while observing this indicator. When a matching point is reached, the SWR light bar drops to the bottom, while the power light bars increase.

fed dipole, but I have never encouraged the use of end-fed antennas). One conductor goes to one side of the dipole and the other conductor goes to the other side.

You may hear many discussions about how long the feed line should be. Simply, however, it should be long enough to reach the shack. A problem arises in that sometimes we end up with *RF* voltages in the shack, usually indicating the antenna system load may be operating at a very high impedance. The very simple answer to this is to change the feed-line length by adding more line. If the type of insulated "open wire" line I mentioned is being used, it is no problem simply to coil a length of the line and insert this new piece at the feed point of the tuner. How much line? A complete change would be a quarter wavelength of line, but the answer is to try different lengths to change the load. A quarter wavelength may not be needed.

Some questions about open-wire feeders bear answering here. For one thing, some amateurs have had problems with Transmatches not being able to handle the antenna system, and they have experienced arcing or some other strange problem. Simply put, a 150 watt Transmatch means exactly that: it is capable of handling only 150 watts, regardless of the load or the system. However, some open-wire-fed dipoles, on some bands, will exhibit high RF voltages that can cause arcs or flashovers. This really means the doggone thing won't handle the 150 watts! I have just discussed how to change this condition by changing the load impedance by altering the length of feed line.

Two Coax Feed Lines

What about using two lengths of coax and only using the inner conductors for a "shielded" open-wire line? (See figure 4-4.) It is possible to do this, *but* several problems may arise. I will attempt to cover them here.

First, one would assume that the impedance of the line using two 50 ohm conductors would be 100 ohms. But that is not important—at least not if we are going to "tune" the antenna system as we would with open wire.

We can connect the inner conductors to the balanced output of the Transmatch in the shack and ground the shields. But what about the shields out at the dipole? Most amateurs who have done this left the shields just "hanging." Similarly, if the center of the dipole is suspended from a tower holding a beam or beams, the shield is grounded to the tower. Figure 4-4 illustrates such a setup.

In figure 4-4 we essentially have two antennas—

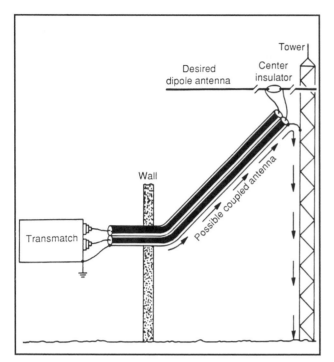

Figure 4-4. A possible "no-no" used by some amateurs: dual coax out to a tower to feed a dipole. The outer shield of the coax from the rig out to the tower, plus the tower itself and the guy wires, can also serve as an antenna. All this could upset a beam pattern if there is a beam (and there usually is) on top of the tower. However, I know some amateurs who swear by such a system, and in that case I go back to McCoy's rule: "If the darn thing works, leave it alone."

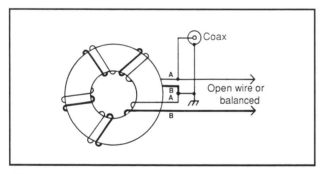

Figure 4-5. This Ruthroff (voltage) balun consists of 17 bifilar (two wires) turns of No. 14 H Thermalize or teflon-covered wire on a T300A-2 powdered iron core for 1500 watts, or on a T-200 core for 200 watts. (For kits contact The Wireman or Amidon Associates.)

that's right, two antennas. The first antenna is a dipole fed by the inner conductors, and the second is an antenna consisting of the tower, from its ground up to the shields and then the shields back to the Transmatch, and its ground. In our world of multi-

bands—from 160 through 10 meters—almost certainly that ground shield antenna is going to be resonant on one of these bands. If so, it will couple power from whichever antenna is being used. If you have such an installation, grid dip the shields to find the bands in which it might be resonant, and change the ground lengths to move the resonance out of the band.

Going back to our dual coax, keep in mind that we still have relatively high-loss feeders, plus these losses shoot up dramatically when high SWR conditions exist. SWRs of 20 to 1 or higher are common on tuned lines with multiband use. The real problem here aside from losses is the high RF voltages that can develop on the lines with the likelihood of line destruction. Frankly, such an installation is just not my cup of tea.

Voltage-Type Balancing Transformers

As mentioned earlier, when I first described the Ultimate Transmatch, I had the problem of going from a circuit that was basically single ended to a balanced output or line (open-wire or twin-lead feeders). The answer boiled down to using a voltage-type balancing transformer. This is important, because over the years arguments have erupted about the ratios, standing-wave matching, and other things. Keep in mind this simple fact: All I wanted was a way to go from unbalanced to balanced line. SWR had nothing to do with the design for a very good reason. If we take the composite antenna system load, the antenna plus feeders, the load presented at the Transmatch can be almost any value, as *we are not matching that load with the balancing transformer.* We are converting this balanced unknown load through the Transmatch to get the load to convert to the unbalanced 50 ohms, the design characteristic of the transmitter or transceiver. Therefore, SWR at the balancing transformer is important only in that the transformer should be able to handle the mismatched currents or voltages.

One other aside here: I mentioned the heating of the current-type balun. The 4 to 1 voltage balun mentioned before, when properly built as described here, just does not "saturate" or overheat, as some people have said. I first made this type of balun in 1957 and have never had problems, and I run the legal limit. It is true that this balun may not be "flat" from 160 through 10 meters, but *we are not using it as a matching transformer.* Our aim is to get from coax to open-wire line and get as balanced a condition as possible. The cores I use are T-200 type material. This material is not as critical as ferrite, which I would not recommend for this application. (Both powdered iron and ferrite cores are available from Amidon Associates and Palomar Engineers, among others.) Again, I refer you to Jerry Sevick's excellent book for more facts on these materials and transformers.

How efficiently can this type of balancing transformer provide balanced output to open-wire line? I know I am being redundant, but the argument that has erupted over current versus voltage baluns leaves one wondering if the whole thing is a tempest in a teapot. In the first place, and I am speaking now about open-wire feed to a multiband antenna, it is well nigh impossible *not* to have some feeder radiation regardless of the type of transformer used. However, it is important to keep in mind that such radiation is not *lost* power. It is going somewhere and probably will work someone. (Obviously, with the current balun tests I mentioned previously, there is power lost as heat). As long as pattern distortion is not a problem, and it rarely is with wire multiband antennas, there should be no concern if there is some feeder radiation.

Since the inception of the Ultimate Transmatch, manufacturers have used the balun/transformer method I described back then in their Transmatch antenna tuners, and I might add with good success. These balun/transformers are normally included as an integral part of the Transmatch and not as separate units. Since this has worked so well for so many years, I am not about to suggest a change. I am always guilty of using cliches, so I guess I can't avoid them. However, believe me, the old cliche that goes "If it ain't broke, don't fix it" certainly applies here! Figure 4-5 is the circuit of this 4 to 1 transformer, and the accompanying photos provide details of construction.

Feed-Line Installations

The fact remains that it is a requirement to get feed lines from your transmitter out to the antenna. There are several methods of using coax, open-wire lines, and Transmatches together. The first method of using coax to get outside the shack to open wire consists of using a balancing transformer outside of the station (not in the Transmatch). We must come out of the transmitter/transceiver using coax to a Transmatch, and then use additional coax out to the balancing transformer, which is capable of converting the system from coaxial line to open-wire line. One important point here: The commonly available built-in antenna tuners in nearly all modern transceivers *will*

not do the job. They simply do not have enough matching range to handle the mismatches encountered with a multiband system.

There are commercial coax to open-wire-line transformers available. However, for those of you who want to roll your own, I will show how to make one in the accompanying figures and photos. That way if something goes wrong and you have to blame someone, that someone will be me. The Wireman sells a kit for making the transformers. As I said earlier, don't be concerned with the 4 to 1 ratio, since we are not concerned with SWR here—at least not in the sense that most amateurs think of SWR, because this is not a matching transformer. Keep in mind that we have a transmitter or transceiver that must work into a *50 ohm load,* and we must provide this load from an unknown antenna system load. The antenna system load may be a very low or very high impedance with lots of reactance, and truly, the balun/transformer is only there to get from an unbalanced to balanced condition—that's all, nothing more. The Transmatch itself will convert the unknown antenna load to 50 ohms. A single T-200 core with Teflon-covered wire is used for powers up to a couple of hundred watts. For 1500 watts, key-down continuous duty, three stacked cores should be used.

Constructing A Balun/Transformer

Construction of the transformer is relatively simple. The T-200 core/cores are first wrapped with a layer of special insulating tape. When using three cores, after all cores are taped, a couple of strips of tape tare used to tie the three cores together.

Next the two free ends of the Teflon wire are put in a vise and the two wires are drawn taut. I suggest using some strips of the tape to fasten the two conductors together, making a two-conductor wire ribbon. Next this wire "ribbon" is inserted through the cores or core and wound until there are at least 10 turns (10 to 12 turns are okay). We now have the transformer. It can be mounted on a metal plate or U, with a coax fitting on one side and two terminals on the other side (for the open-wire line). The entire unit can then be mounted on an outside wall, a post, or whatever (see figure 4-6). When mounted outside the assembly should be covered with a plastic freezer box to keep out the rain and snow. The Wireman tells me that they can supply two different kits—high or low power.

As I noted earlier, there are a couple of commercial units sold in which the transformer is already mounted

Here are the three powdered iron cores wound with 3M insulating tape. These are the ones I used to construct the 4 to 1 transformer/balun shown in figure 4-5.

in a case. One in particular is the MFJ-912, W9INN remote transformer, which is identical in circuit to the transformer I described above. This unit uses two cores, but they are larger than the three I specified in the home-built version. They will handle at least 2 kilowatts with mismatches that are very high—10 to 1 or greater. This unit is already in a weatherproof box with mounting screws and terminals.

The Importance of Coax Quality

Remember, it is important that the coax from the Transmatch in the shack out to the transformer be a high-quality line such as an RG-8/U type, which will handle 4000 to 5000 RF volts. Possibly RG-8/X can be used for the 150 watt level, but I would recommend heavier coax to be sure. To repeat: The reason

for caution here is that a high SWR can exist on this line between the transformer and Transmatch. A high SWR can cause high voltage or high current to be present, which could cause damage to the coax line. For this reason it is best to keep this coax line as short as possible.

I realize I did not use a specific length here, but short as possible means just that. I know of some amateurs who have used as much as 50 feet of coax, but I think that is begging for trouble. I personally go to great lengths to ensure that my lengths are under 25 feet because a high SWR is likely to exist on this coax. For example, an 80 meter dipole, 130 feet long, center fed with open-wire line is likely to have an SWR of 10 to 1 or higher appearing on the line. This mismatch also appears across the transformer and then goes on to the 50 ohm coax. The Transmatch in

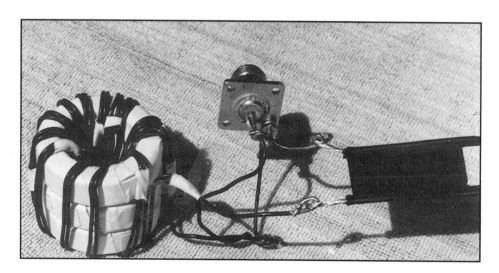

I have deliberately left this as open-type construction to show the bifilar windings and the connections to the coax fitting (SO-239 type) and the other two leads connected to the open-wire lead. While electrically this would work well, in actual practice the transformer needs to be mounted in a weatherproof box or case. Details of the installation are included in the text.

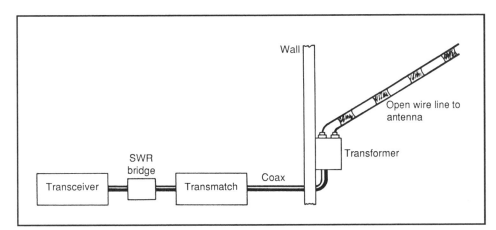

Figure 4-6. This drawing demonstrates our goal, as described in the text, of bringing a coax through a wall (or whatever) out to a transformer. The output of the transformer connects to the low-loss open-wire line.

the station converts the mismatch to a pure 50 ohm match. However, there still can be high voltages and currents on this coax, so keep it as short as possible.

Tune up has been covered before, but it is so simple it may warrant repetition. With the system all connected and an SWR indicator in the line between the transmitter and Transmatch, enough power should be applied to obtain a reflected reading on the SWR bridge. The Transmatch is then adjusted for a null or match as indicated by the SWR bridge. The power is then brought up to the desired level.

Installation Without A Balun/Transformer

The second installation system does not require an external balun/transformer at all, but uses the one which is built into the Transmatch. Use two equal lengths of coax to get from the balanced-line output of the Transmatch to the open-wire line outside the house (see figure 4-7). The coax is run together, and the two inner conductors will be the feed line. The coax shields are connected together at each end (soldered). The coax feeders at the Transmatch end are connected to the balanced output terminals on the Transmatch. (In this case the balun/transformer is already built into the Transmatch; just about all commercial units have them). The coax leads at the outside point are connected (soldered) to the open-wire conductors, and the shields are connected to earth ground.

Wait! I know what you are going to say. What about the impedance of the open-wire feeders (450 ohms) being connected to the parallel coaxial feeders (100 ohms)? Isn't this a bad mismatch? It is a mismatch, but it is of no importance in this case. The coaxial section is a balanced line, and there can be no radia-

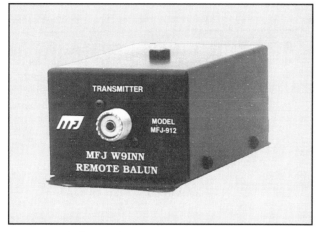

This is the MFJ-912 four to one open-wire to coax transformer. A rugged unit designed to handle large mis-matches.

tion from it. (Not only that, it is shielded.) The difference in line impedance does not matter, because we are matching or adjusting a *complete* and complex antenna system load via the Transmatch.

We could, if we wanted to, make a transmission line of combination impedances—say, 300 ohm line, 450 ohm line, 600 ohm line, etc.—as long as both conductors in each line are perfectly equal in length. In theory, the radiation from one conductor cancels the radiation from the other, so the line doesn't radiate. If that is true, the only problem is one of matching this completely unknown load back to 50 ohms at the transmitter—and that is what we do. We adjust the Transmatch as described earlier, keeping the dual coax lines as short as feasible. The same or similar high voltage/currents mentioned earlier can exist here.

While I have written here about what we can do, it is not what I do. I don't care if the open-wire line comes into the shack. Although I have thoroughly tested both

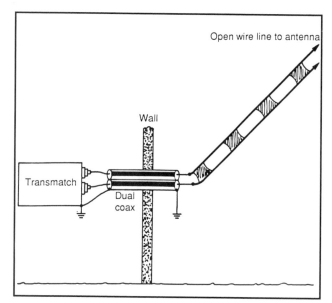

Figure 4-7. Here is the second method of using short lengths of coax to get outside the shack. The dual lengths of parallel coax are brought from the balanced outputs of the Transmatch out through the wall. At that point they are connected to open-wire low-loss line, which feeds the antenna.

methods, as I stated at the outset, I bring my open-wire line in along with several coax lines through a section of PVC pipe and then up to the Transmatch.

An Alternate Approach

I know of one amateur who replaced a pane of glass in a window near his shack with a sheet of clear, heavy

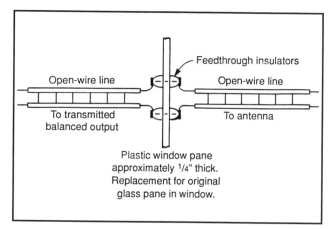

Figure 4.8. Shown here is another method of feeding open-wire line directly to the antenna. A window pane is replaced with clear $1/4$ inch Plexiglas^{TM}. Two standard feed-throughs provide a means to interconnect open-wire line from the inside directly to the outside. Plastic pane can also be used to mount a coax feed-through, such as UG-363, to interconnect coax feeds in the same manner.

plastic ($1/4$ inch thick). He installed two feed-through insulators $5/8$ inch apart to connect the inside and outside lines together. This installation worked effectively for many years with no problems (see figure 4-8). This same plastic window pane was also capable of accepting a coax feed-through, such as UG-363. This eliminated the requirement to drill holes in the wall to get coax outdoors from the rig. If you try it, be sure to support the coax properly on both sides of the plastic window pane to reduce any stress.

Chapter 5

Wire Antennas

This brings us to antennas—what is good and what is better. I grew up with wire dipoles, and in fact with many wire beams. Most amateurs starting out on the low bands (160 through 10 meters) immediately look for an "all-band antenna." There are many wire "all band" antennas made commercially, or you can make your own.

As was pointed out earlier, modern transceivers will not work if the SWR developed by the load, or the impedance, exceeds 2 to 1, because the transceiver reduces power or shuts down. No matter how the antenna is fed, it isn't always simple to stay below this 2 to 1 figure. Some amateurs frown on the use of a Transmatch because it means extra controls to adjust. However, because of modern SWR indicators, the adjustments are very simple (and are discussed in detail in chapter 4). But while some amateurs don't like using a Transmatch, there are some basic facts that must be considered.

In the first place, nearly all (in fact, I know of none that do not) transceivers and amplifiers are designed to work into a 50 ohm impedance or load. Any, and that means *any*, deviation from a 50 ohm load means that the amplifier stage is not going to work properly and will be inefficient. There has always been tremendous pressure applied to antenna manufacturers to produce antennas that present 50 ohms over a broad range of frequencies. However, the facts of life relating to antenna design are not that simple. It is well nigh impossible to design antennas that will present a 50 ohm load on all frequencies.

Let's assume for a moment that we wish to have the 80 meter dipole described earlier cut for the center of the band, coax fed (50 ohms), and we want to work both the high and low ends of the band (see figure 5-1). Obviously, it is impossible to do this with modern

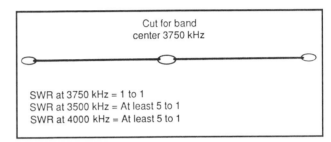

Figure 5-1. If an 80 meter dipole is cut for the center of the band and fed with 50 ohm line, the resulting SWR is going to be very high at either end of the band.

transceivers simply because they shut down above a 2 to 1 mismatch. An 80 meter dipole fed with coax is only going to be less than 2 to 1 over a rather narrow portion of the 80 meter band. This dipole will provide a good match at resonance in band center, but is likely to go up to 10 to 1 at the band ends. Suppose we decide to accept 10 to 1 losses because of mismatches at the band ends but still want to use coax. We will have to go to a Transmatch to convert that 10 to 1 load to a 50 ohm transformed match.

Will our coax take high power with the legal limit of 1500 watts going through the line and a 10 to 1 mismatch? Remember earlier I mentioned I would avoid discussing voltage and currents. In this case, however, we have to look at voltage ratings. The SWR is the ratio of RF voltages (or currents) at any point in the line. With the normal coax used in a legal-limit station (for example, Times Wire and Cable RG-8/U), the coax has a maximum rating of 4000 volts. This may seem like a very high rating, but high SWRs, such as 10 to 1, can create some very high RF voltages which exceed the rating of the coax, resulting in destruction of the line. The fundamental point here is that these limits should be considered when attempting to use coax as a "tuned" line with a Transmatch and high SWR.

If you understand the basic concepts presented thus far, then you must realize that if open-wire line is lossless, then antenna impedance is not a consideration. And that is absolutely correct! On the air you may hear many amateurs state that an antenna will only operate best if it is a resonant antenna. This is pure hogwash, which requires some explanation.

The Importance of Eliminating Reactance

If an antenna is resonant, there is no reactance present. Therefore, most of the power fed to the feed point will be accepted by the antenna and radiated. But if an antenna is not resonant, we can still feed all the power you generate into the antenna. If we cancel out all the reactance present, power will be radiated—regardless of the electrical length of the antenna. Once you realize this simple point, antennas—particularly multiband antennas—can be understood.

From what you have already read and learned here, you must realize that *any* metal conductor can be a multiband antenna. To repeat and emphasize, *anything that will conduct RF can be an antenna, even a multiband antenna.* Strange as it seems, a simple paper clip could be a multiband antenna assuming we could get the RF power into it. It would be terribly inefficient, because the ratio of ohmic losses to radiation resistance would be astronomical. More on this later when we discuss mobile antennas.

The Antenna System

If any wire is a multiband antenna, what is the catch? Obviously, we need to match a multiband antenna over the frequency ranges of interest to prevent the existence of very high SWRs. This leads us to the use of a low-loss feed line to prevent power loss because of a high SWR—such as open-wire line, which is an essentially lossless line. In addition, we will need a wide-range adjustable circuit—a Transmatch.

Now comes an important point which was addressed briefly in the last chapter. We will not be matching an antenna, but rather an antenna *system.* The antenna, the feed line, and the Transmatch all become a system, and our simple problem is that of converting the unknown load at the transmitter end of the feed line into a 1 to 1 match, or *a resonant system for the frequency we plan to use.*

We need to deviate here again and explain another point. Simply because a feed line has a given impedance, as explained earlier, does not mean that the "load" presented to the Transmatch is going to be the same impedance as the line. It probably won't be, unless the line is matched to the antenna impedance. There is likely to be a great deal of reactance present at the Transmatch end of the line. Our job is to convert/transform this unknown load to match the transmitter final amplifier design output—customarily, 50 ohms, non-reactive. And that is exactly what we do.

Multiband Dipoles

Of course, as I pointed out a few times already, a multiband dipole can be any reasonable length and can be used on all bands. What is a reasonable length? The McCoy dipole is simply a dipole that fits between two supports. I recommend that the *shortest* length used be at least a quarter wavelength long at the lowest operating frequency, which is 65 feet if the antenna is to be used on 80 meters. That is not to say the antenna will not perform on 160. It will, but not as well as a longer antenna. Just make your wire long enough to fit between the two highest supports you can find. Cut it in half (that makes it a dipole) and feed it in the center with open-wire line. Of course, you can make it an inverted V by supporting it at the center and getting the ends as high as possible. That is a good, even an excellent, multiband antenna which can be built by the user at very low cost. No advertised, commercial wire multiband antenna will outperform this antenna.

Don't misunderstand. Some amateurs simply do not want to "roll their own" and prefer to buy. There are many good multiband trap dipoles, or Windoms, available. I want to emphasize, though, that if you want to build your own, the antenna described above is excellent.

In this day and age it is impossible to ignore some good multiband (or single band) antennas that are commercially made.

In chapter 12, mobile antennas, I will describe a multiband, 80 through 10, that is essentially homemade. A commercial version of that same antenna is made by High Sierra Antennas, and it is multiband. This antenna is basically a vertical, or as you will see in the chapter on mobile antennas, both a horizontal and vertical antenna! In any case, the antenna doesn't need a Transmatch. It is coax fed, and can always be set to an SWR of 1 to 1. Sound like magic? Well, it isn't. Just good electrical engineering by Don Johnson, who will be discussed in detail later.

Chapter 6

Feed-Line Radiation Problems

The purpose of a feed line—coax, open wire, or whatever—is to transfer power from the rig to the antenna as efficiently as possible, preferably without radiating. If it radiates, it is no longer a feed line. It is an antenna. However, the following premise must be stated: Feed-line radiation is not necessarily a bad thing. It is only bad when it upsets the desired radiation pattern from a directional antenna, such as a beam. In wire multiband dipoles or even single-band dipoles such radiation is not "lost" power. This radiated energy will go somewhere and possibly work someone for you.

Any flow of current on the outside of a coax feed line is known as "parallel standing waves." This current can also flow into your SWR bridge, and by combining with the normal SWR readings can give false readings. If different SWR values are measured as the bridge is moved to different places in the feed line, then aside from normal line losses, such changes in readings mean parallel standing waves probably exist.

RF Flow in Coax

Let's discuss RF flow in coax. The RF generated in a rig flows from the rig to the antenna in the two-conductor coax. The two conductors are the inside, or center, conductor and the shield that surrounds the inner conductor. One of the toughest things for amateurs to understand is how coax can be a two-conductor line when that outer shield is grounded at the transmitter. The explanation of this phenomenon is very technical, and I promised not to be too technical in this book, so in this case please accept a "simple" explanation. This is a case where I must again repeat points to try to drive the lesson home.

The actual flow of the RF takes place on the *outside* of the *inner* conductor and the *inside* of the outer shield. Regardless of the fact that we connect both the inside and outside of the outer shield to one side of the antenna, there should be no flow on the outside of the outer shield. All of the RF flow takes place inside of the coax (see figure 6-1). That being the case, we can ground the outside of the outer shield without worrying about RF being there. Ahhhh . . . if life only were that simple! Unfortunately, when the feed line reaches the antenna, we must attach that outer shield to one side of the dipole or other antenna. If the outside of the outer shield happens to be a resonant length of wire, from the antenna down to earth ground, then we have a problem. A basic rule of radio

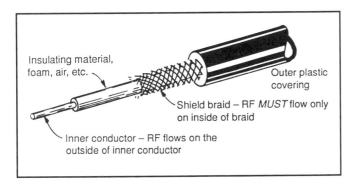

Insulating material, foam, air, etc.

Outer plastic covering

Shield braid – RF *MUST* flow only on inside of braid

Inner conductor – RF flows on the outside of inner conductor

Figure 6-1. A drawing of a section of coax with the insulation stripped back. The coax has an insulating outer cover, which, depending on the grade of coax, can protect the inner conductors, even to the extent of being buried for up to 20 years! RF flow, as mentioned in the text, must be on the inside of the outer braid and on the outside of the inner conductor.

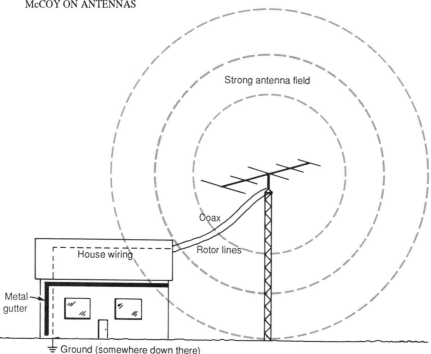

Figure 6-2. An antenna installation may be surrounded by many metal items that can be resonant in an RF field and cause problems. In this drawing we have rain gutters, house wiring, and, while not shown, power lines and pole ground leads, all of which can cause problems.

If ANY metal around antenna field is resonant, that metal becomes an antenna and can reradiate. This may or may not cause a problem.

and antennas—and never forget this—is if a length of metal in the antenna field of energy happens to be resonant to the energy field, then RF is going to be coupled to that metal (see figure 6-2).

This essentially means that if the outer RF shield is resonant to the frequency in use, RF will be coupled to that shield, flow on it, and be reradiated from it. In plain language, the shield has become an antenna. If you think about this for a moment you will quickly realize that any telephone lead, or TV antenna lead, etc., that is in the vicinity of the antenna field can also become a problem! But right now, we are not interested in interference possibilities, only what happens to feeder radiation.

Keep the aforementioned point in mind as I try to lead you out of the darkness. Any resonant length of line will couple energy from the antenna field. Some years ago the following theory was advanced: Because coax is an unbalanced transmission line used to feed a balanced antenna, a balun could be used to cure problems that might arise, such as coax shield radiation. A balun is simply a transformer that converts the "unbalanced" coax to the balanced load of the antenna. (Chapter 4 covered this balun application in detail.) Installing a balun on an antenna or installing ferrite beads on a feed line may or *may not* cure the coax shield radiation. In fact, it could actually create a problem if it alters the length of the outside of the outer shield, making the shield resonant where it

wasn't before. Think about it. That is not the answer, however. Be aware from figure 6-2 that even if we successfully avoid or correct feed-line radiation, there still may be other metal objects around a station that can cause re-radiation and beam-pattern destruction. Note the similarity of the radiation problems of figure 6-2 and figure 4-4.

Now go back to what you have read about resonance, which we defined as the absence of reactance. Remember I stated that reactance can be described as a gate or door that stops or reduces the flow of RF. If we are careful, we can eliminate any outside shield radiation by making sure that the outside of the shield is not resonant. In brief, it is very difficult to couple power to a non-resonant circuit because of the reactance present.

Determining if a coax shield is resonant is rather simple. First we beg, borrow, or whatever a grid-dip meter and make a small loop in the coax feed line to couple the grid-dip coil tightly to the line (see figure 6-3). Then we tune through the frequency bands for which our antenna is designed, looking for a dip on the grid-dip meter. If we find one, we know the shield is resonant and will re-radiate power. It is important to understand that the *electrical* resonant length will be the length of the shield to the point where the line is returned to earth ground. It is also important to realize that we do not know where electrical ground actually is. It could be at the earth's surface or many feet down, but the grid-dip method will show resonance regardless. Next we add some line (we don't prune line, as the stuff is too expensive!) to move the resonance out of the band or bands. We only have to go a few hundred kHz out of a band to get rid of the problem.

RF Chokes

One of the popular so-called "baluns" is the installation of RF beads on the circumference of a coax at the antenna or transmitter end. This is not really a balun, but an RF choke which stops or "chokes" the flow of RF on the shield. This will help cool off lines in many cases, but still won't be too effective if the outside of the coax shield happens to be resonant. I have never really liked the outside bead method, because of tests I conducted using beads applied to coax of resonant lengths fed by both low and high power. The beads became warm at low power and actually shattered at high power. This confirmed that I was wasting power heating the beads! Power converted to heat is lost power! Remember what I wrote earlier: A one-half dB here and a one-half dB there is very important! The best approach to cure any problem is to eliminate the problem. The grid-dip method always works; baluns work sometimes. Put on your thinking cap. If the outside of the line is not resonant, no RF. If you make the outside of the line non-resonant, it is certain to cool off and work effectively.

Many amateurs have used quarter-wave chokes at their antennas by making a quarter wavelength of just the coax shield slipped over the coax at the antenna. The antenna then "sees" a very high reactance at the shield and doesn't couple power back from the antenna. The same result is achieved by making the outside coax line non-resonant; once it becomes reactive there is no need to add a choke!

Common-Sense Installations

As I previously pointed out, feeder radiation is not normally a serious consideration in multiband wire antennas, because the radiation is going to go somewhere and possibly help make a desired contact. However, this essentially is a case of using common sense to determine if feeder radiation will be a problem.

One of the dumbest things I have ever seen in amateur radio is the use of baluns at the feed point of 160 and 80 meter antennas. Amateurs will state that they don't want to "upset" the patterns of these dipoles, so they install baluns (which can be expensive). Why is

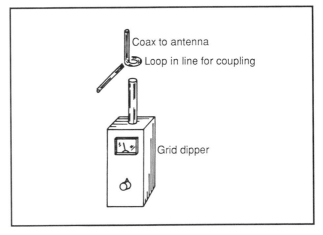

Figure 6-3. In order to get tight enough coupling, we suggest making a tight loop in the coax and inserting the grid dip coil inside the loop. This will ensure good readings.

it dumb? Very simple. In order to achieve a reasonable pattern from dipoles on 80 or 160 meters, the antenna has to be very high above earth. By high I mean at least 150 feet on 80 meters and much more than that on 160. Antennas at lower elevations radiate the majority of their power (90 percent) straight up. Useful low angles only occur at heights of a wavelength or more. Feed-line radiation is absolutely of no consequence. In fact, such radiation may produce some useful low-angle radiation. For this reason, putting in a balun is pointless!

Where is feed-line radiation important and what is "common sense" here? Let's assume we live in a five-story condominium, and we have permission to put an antenna on the roof. The feed line must run up a duct or elevator shaft or along an outside wall. Stop and think for a moment. The feed line is going to be passing close to TV sets, telephones, and who knows what else. All of these devices are susceptible to interference caused by near fields of RF. I know you are going to say, "Come on, McCoy. Use coax feed, which is shielded!" I have just pointed out that RF can flow on the outside of coax and radiate, so obviously the answer is not just to use coax. However, an answer provided a little later in this book is the use of the Spencer antenna method. In the Spencer method dual coax is used as a feed line to eliminate some of the problems inherent in connecting the shield to the antenna. In the Spencer method the outside of the shield is very unlikely to become an antenna.

In beam installations we certainly don't want feeder radiation. Years ago two amateurs bought identical towers, identical beams, etc. They noted that one had poor front to back while the other had good front to back. The figure variation from band to band for identical commercially manufactured beams just didn't make sense. I happened to become involved, and the three of us worked for several days trying to analyze the problem. We even switched beams, because we thought it might have something to do with location. However, the performance in each location remained unchanged.

Finally, I got the bright idea that maybe—just maybe—the feed-line shield was radiating. After much thought and discussion we decided to use a grid-dip meter. We found that one station had feedline resonance on 15 meters while the other did not. In addition, the second station had resonance on other bands, whereas the first one did not. We changed the lengths of both feed lines to get the resonances out of all bands. You guessed it. Both beams now had exactly the same front to back, etc.

The above tests were made before the advent of baluns. Since then I have made many more tests, sometimes adding baluns, sometimes removing them. I discovered that adding a balun actually could put the shield length in a resonant condition. This convinced me to avoid baluns (except for the applications discussed earlier). However, I would be very foolish if I didn't acknowledge that some amateurs insist on baluns and they would be very unhappy without them. To use another cliche, you pays your money and takes your choice!

Let's refer back to chapter 4 for a moment. My reference in the above paragraph to earlier discussions goes back to my use of 4 to 1 transformers/baluns to go from balanced to unbalanced lines. There also are times when a balun is needed to get from one impedance to another. Again I refer you to Jerry Sevick's new book, *Building and Using Baluns and Ununs*, for appropriate details.

Chapter 7

Some Basic Antennas

The standard single-band antenna used by many amateurs is a coax-fed, simple half-wavelength dipole, as shown in figure 7-1(B). An open-wire-fed half-wave dipole is shown in figure 7-1(A). A dipole antenna consists of two equal-length conductors fed in the center. The length of a half-wavelength dipole is determined by the formula mentioned earlier—468 divided by the frequency in megaHertz. There are two common configurations for a dipole—either a horizontal antenna (best) or an inverted V. Both versions are shown in figure 7-1.

Antenna Efficiency

It should be pointed out here that there is another factor controlling the efficiency, or gain, of one antenna when compared to another. This is known as the *effective aperture* of an antenna (sometimes incorrectly referred to as "capture area"). A horizontally mounted dipole has a larger effective aperture than an inverted V, and hence has slightly more gain in its best direction. This difference is very slight though.

In many cases an amateur is restricted to a single tower or pole, so the inverted V is preferred simply because the main support for the inverted V is the tower or pole. Over the years there have been many arguments as to the best angle to use for the legs of the inverted V. It seems that 45 degrees on each leg is preferred. However, a simple fact is the horizontal dipole is definitely better, so it would appear to me that the closer we can bring the end supports up to the horizontal, the better.

Dipole Characteristics And Considerations

As pointed out earlier, the impedance of the dipole will depend a great deal on its height above ground. In a normal installation—if there is such a thing—the impedance of the dipole is approximately 20 to 100 ohms (usually closer to 70 ohms). Coaxial cable of the 50 ohm variety should prove a good match with an SWR of less than 2 to 1.

As to antenna wire size, the only consideration worth thinking about is strength. For example, in a

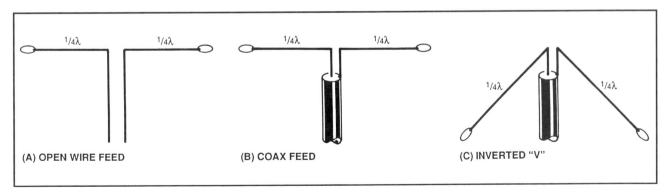

(A) OPEN WIRE FEED   **(B) COAX FEED**   **(C) INVERTED "V"**

Figure 7-1. At (A) is the basic dipole, and at (B) is a dipole as a half-wavelength antenna. Each side wire is one-quarter wavelength long. Probably the most popular configuration is shown at (C), where the antenna is in the form of an inverted V. As pointed out in the text, an inverted V is never quite as good a performer as a horizontal dipole.

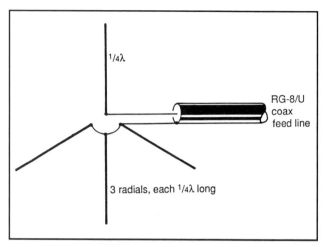

1/4λ

RG-8/U
coax
feed line

3 radials, each 1/4λ long

Figure 7-2. The more or less standard ground plane antenna. The reason for the "more or less" is simply that the ground side, instead of being only three radials can be many more (up to 30 or more). If, for example, we were going to build a ground-mounted vertical for 40 meters, the vertical would be approximately 32 feet high (a quarter wavelength on 40) with at least 30 to 50 radials, each one of which would be no less than two tenths of a wavelength long—preferably 32 feet long. The radials should not be buried more than 1 inch down. The impedance of such an antenna would be on the order of 37 ohms, suitable for direct feed with 50 ohm coax.

center-fed antenna that is suspended by the ends, a strong wire will be needed to support the coax line. As an inverted V, where the feed point is supported on the pole or tower, a smaller size antenna wire can be used. Keep in mind that for all practical purposes No. 18 copper-weld wire is just as efficient a radiator as No. 12. Many amateurs have used wire as small as No. 32, supported with rubber bands(!), to make an "invisible" antenna. High power can be run to these fine wires as long as they are not too heavily insulated.

Questions often asked are "Is it possible to drop the ends of an antenna to increase its overall length?" and "Can the ends be run in a Z pattern to get more length in a smaller space, as in an attic?" The answer to both questions is yes, but never expect a "confined" or smaller space dipole antenna to be as efficient as a full-size, stretched-out dipole. The effective aperture gets into the act, and as an antenna is fitted into a smaller space, its performance suffers.

Dipole Radiation Patterns

The radiation pattern of dipoles always raises questions. One of the most common mistakes amateurs

make occurs on 80 meters, or even on 40, by not understanding radiation patterns. For example, a dipole has a free-space pattern of a figure-8. However, we are not in free space, and the controlling factor that applies is the antenna's height above ground. For example, I would venture to say that the average amateur gets his 80 meter dipole, either horizontal or inverted V, up somewhere between 30 and 50 feet high. He goes to great lengths to orient the dipole so that the theoretical figure-8 radiates in his desired direction. Unfortunately, an 80 meter dipole lower than the optimum height above earth has an omni-directional pattern, with most of the energy being radiated straight up! Such a pattern is great for local work on 80 meters, but isn't too effective for DX. So don't worry about antenna orientation on 80 or 40 unless you can get the antenna up at least one-half wavelength (130 feet or so on 80). Don't give up hope, however. In chapter 8 I'll show you an 80 meter loop that will produce effective DX angles, and the antenna doesn't have to be more than 30 to 50 feet high.

The high-angle radiation problem only applies to 80 and 40 meters. When we start thinking in terms of the higher bands—20 meters and up—a dipole will produce a DX angle at much lower heights. For example, a 20 meter dipole 30 feet above ground will produce useful angles of radiation for working DX. And because rotary beams produce gain by concentrating energy radiation, a rotary dipole is not to be scorned.

A very popular antenna I described many years ago is a 15 meter rotary dipole made from electrician's thin-wall conduit and mounted on a wood 2×2. It is described in chapter 10. The antenna is very inexpensive, costing less than $15 to construct. This rotary concentrates the figure-8 lobes in the desired direction and provides excellent front-to-side rejection.

Dipoles Compared To Beams

A dipole is considered to be a basic antenna, and the gain of beams is usually compared to the performance of a dipole. For example, a three-element beam will have approximately 7 dBd forward gain compared to a dipole. Plus the beam will probably have about 20 dB or so front-to-back rejection. I like to convert such numbers to a real-world, easy-to-understand language. A simple comparison of forward gain using a receiver calibrated at 6 dB per S unit is the gain of a one-element rotary beam will be approximately one S unit weaker than that of a three-element beam. In many instances, although it goes against my statement of the

This is a random-length, end-fed wire (about 35 feet long) that I used on my trailer. The antenna was used on 80 through 10 meters with the use of a Transmatch.

importance of a one-half dB here and a one-half dB there, it still is a very respectable signal.

Monoband Verticals

Most of my discussions so far have been aimed at horizontal antennas. I would be remiss if I didn't include some information on monoband verticals. I promised no formulas except the one (468/F). I won't give you the formula for a quarter-wavelength vertical except to say use the 468/F formula and divide the result by two; the answer is the length of a vertical. For example, 16 feet 2 inches produces a half-wave-length dipole for 10 meters. A vertical for the same frequency would be one half that, or 8 feet 1 inch.

The impedance of a half-wave dipole is on the order of 70 ohms, and the impedance of a quarter-wave vertical is just about half that, or 35 ohms. Without getting too technical, verticals need a ground to work against, and usually we think in terms of a vertical using a ground plane. Figure 7-2 illustrates a ground-plane vertical. I have shown three ground-plane radials, but the more radials you install, the better. If the radials are run out horizontally, the impedance of the vertical, using quarter-wave lengths, will be approximately 35 ohms or so. If the three radials droop down about 45 degrees, the impedance will approach 50 ohms, making the antenna a direct match for 50 ohm cable. (Of course, if the radials are drooped straight down, the antenna becomes a vertical center-fed

A really inexpensive 2 meter one-quarter wavelength vertical. Cost is zero when using a wire coat hanger.

halfwave, and the impedance will approach 70 ohms.)

As to radiation patterns, the vertical will produce a much lower lobe(s) than a horizontal antenna. A 65 foot high metal pipe or tower would make an excellent low-angle 80 meter antenna, but 65 feet of metal pipe straight up is not easy to come by.

VHF and UHF Antennas

I hadn't planned to go into much depth on the subject of VHF and UHF antennas in this book. However, one

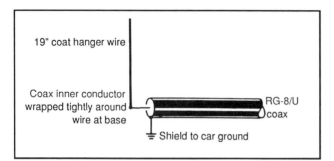

Figure 7-3. This is the simple, 19 inch makeshift vertical I used on my van. It worked very well for something that didn't cost anything.

little inexpensive gem I came up with is worth describing and showing.

I use a 2 meter $^5/_8$ whip mounted on the trailer of my RV. On one trip the top of my trailer brushed some tree limbs and I lost the 2 meter vertical. I did some pondering and decided to make the cheapest and easiest vertical for 2 meters one can find. I straightened a metal coat hanger and cut it so that I had 19 inches of stiff wire, 19 inches being a quarter wavelength on 2 meters. I scraped off the enamel covering on one end and then skinned back some coax so that I had a couple of inches of inner and outer shield exposed. I tried soldering the inner conductor to the end of the 19 inches of wire, but couldn't get enough heat to do the job properly. I therefore wound the inner conductor as tightly as possible around the end of the wire and securely taped it. I then taped the antenna to a piece of phenolic plumber's tubing and mounted the assembly

on my Suburban roof, with the shield connected to an adjacent screw. An accompanying photo shows this sloppy installation. While it was sloppy, it got me on the air and did an outstanding job. I felt as proud of that installation as I have of some of my more exotic beams! Figure 7-3 is the electrical drawing.

I had such good luck with the antenna that I thought I might try an end-fed wire for the low bands, the wire being mounted over the roof of my trailer. The wire was made as long as possible. It came out to about 37 feet and was end fed with a Transmatch. Another accompanying photo shows that antenna. The antenna was a 40 meter or so quarter wavelength, and while it worked well on 40, it did even better on the higher bands. I suppose it therefore should be in the multiband antenna chapter.

DX Halfsquares

At one of the recent amateur radio shows Jim Stevens, KK7C, of Antennas West, spoke on antennas in general, including some that he manufactures. I was very impressed with Jim's practical knowledge and with an antenna he called his DX Halfsquare. Since this book presents what I think are good antennas, particularly those of low cost, I have no qualms about including some of Jim's gems in this book.

Jim's antennas are designed as monobanders, and each halfsquare consists of two vertical quarter waves connected together by a half-wave horizontal line. Figure 7-4 shows the antenna configuration. Radiation is essentially broadside to the array with a rather low

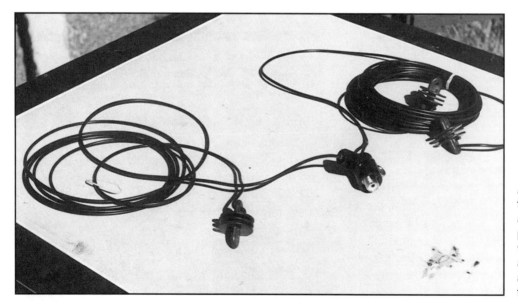

A half-square antenna is furnished with insulators and a feed terminal. A poor-man's DX setup would be a couple of these antennas, some A-frames, and the antennas oriented for the best DX path.

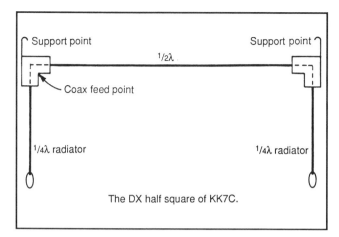

The DX half square of KK7C.

Figure 7-4. The commercial half-square antenna sold by Antennas West (KK7C). This antenna consists of two quarter-wave sections separated by one-half wavelength. Directional radiation is essentially broadside. The gain is on the order of 4 to 5 dB and makes for a low-cost, low-angle DX antenna.

effective DX angle. While the halfsquare is essentially a monoband antenna, a Transmatch can be used to tune it as a random-fed wire.

The halfsquare has an approximate gain of 4 to 5 decibels broadside to the array. An important point is the main angle of radiation is low, making the antenna very effective for DX. I have tested two different half-square antennas on two different frequencies and found performance to be equal to my three-element beam.

An Inexpensive Antenna Support

An antenna support that has been used effectively for a very long time is the A-frame type of mast. The A-frame (figure 7-5) consists of three lengths of wood 2×2's. I have made A-frames that were 30 feet high and have heard of others much higher. In any event, the drawing provides all the information needed. A pulley and line are mounted at the top of the A-frame to raise and lower antennas. Three guy lines are needed at the top and three at the center to support the

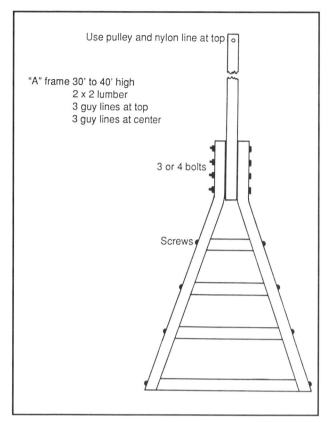

Figure 7-5. The A-frame, an amateur's standby for putting up antennas. The A-frame has been around since the beginning of amateur radio. It is easy to assemble and can be used (with great care) at a height of up to 40 feet. It usually takes at least three people to "walk" up a 40 footer—one to hold the base, another to keep the top guy lines taut while it is going up, and a third to do the lifting.

mast. I have had A-frames stay up for years, so I know they are reliable and inexpensive.

Such a support for an inverted V can be used. One word of caution about inverted V's or any dipole: The high-voltage points are at the ends of the dipoles. If you come in contact with the ends when there is RF present in the antenna, you can get a nasty RF burn. Therefore, the ends of the antenna should be mounted high enough to be out of reach.

Chapter 8

Multiband Antennas

This chapter will cover multiband dipoles, long wires, loops, verticals, and more. Let's begin with multiband dipoles.

Multiband Dipoles

There are several homebrew multiband dipoles that can be fed with either coax or open-wire/ladder line. Again, the question to be asked is "Can a homemade multiband dipole be as good as or better than a commercial unit?" The answer is definitely yes! Not only can you make your own multiband antenna, but a lot of money can be saved in the process.

As I already have pointed out, without a doubt one of the best multiband antennas is simply a dipole as long as possible and as high as possible. Such an antenna using open-wire feed line of the insulated type plus a Transmatch can cover all of the amateur bands from 160 through 10. But suppose someone insists on using coax feed. Let's assume that this individual wants to cover 160 through 10 meters as efficiently as possible.

You have learned by now that anything that conducts RF can be a multiband antenna. You have also learned that very, very short antennas can have high losses. You have learned that any dipole fed with a low-loss line such as open wire can be used via a Transmatch to present a matched load, 1 to 1, on any band or frequency. Our first question logically should be "How short can I make my multiband antenna and still have it be a good performer?"

The general rule that has been followed over the years is one should shoot for a length that is at least one-quarter wavelength long on the lowest band. In other words, a dipole that is at least 60 feet long will do a fairly good job on 80, and of course, will do even better on the higher bands.

I have been asked many times what I use. Years ago at an antenna lecture in California I was asked how I made my multiband antenna. I told the group that I had found two supports—one a tower near my shack and

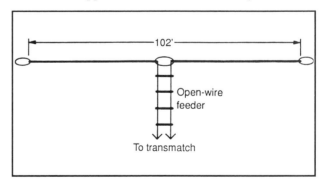

Figure 8-2. This is the G5RV antenna as originally published by G5RV. The real secret of the G5RV is that it is long enough to be a good performer on the lowest band (80 meters). Also, the feed impedance is close enough so that minimum matching or tuning is required (a very broad technical statement!). However, as pointed out several times in this book, feed impedance depends a great deal on each installation for this or any other antenna.

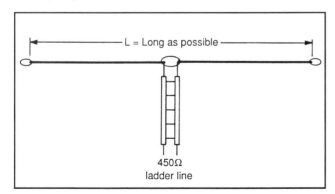

Figure 8-1. The McCoy Dipole is made as long as possible, trying to keep it longer than one-quarter wavelength overall. The antenna should be fed with 450 ohm ladder line (it is inexpensive and very good). (See text for details.)

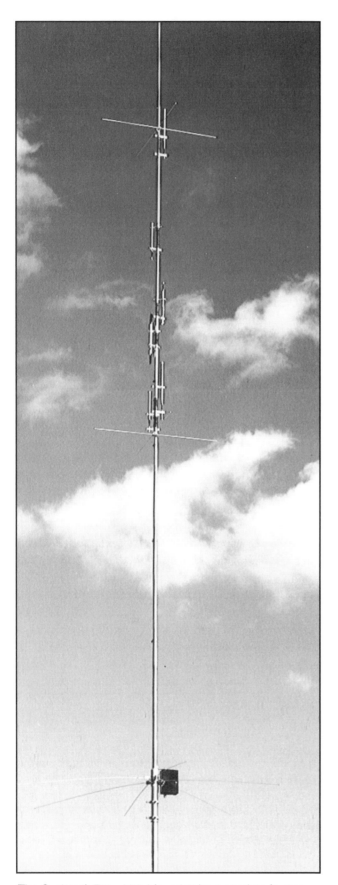

The Cushcraft R7, which I found to be a good performer.

the other a pole (power) near the rear of my property. I then cut a wire that was long enough to fit between the supports. Did I measure the wire? No! I cut the wire at the exact center, used an insulator, and soldered on open-wire feeders. The feeders were long enough to reach my station Transmatch. What I had was an antenna that was long enough to fit between the two supports. I later measured the dipole and found that it was 164 feet long (see figure 8-1). The antenna was longer than an 80 meter dipole, but shorter than a 160 meter half wavelength. Someone in the audience asked me what this particular antenna was called. I couldn't resist, and simply said "a McCoy dipole." The name stuck. Any dipole of an "odd-ball" length is a McCoy—hi! The exception is if the antenna is 102 feet long, then it must be called a G5RV antenna.

The G5RV (Varney) Antenna

Some years ago R. L. Varney, G5RV, found that 102 feet, center fed, worked out to be a reasonable match for 50 ohms on 20 meters when used with a matching section of 300 ohm line. Amateurs soon discovered that under certain conditions the antenna would also provide a reasonable match on other bands. The 102 foot length therefore has become popular and is known as the G5RV antenna. The actual and original G5RV is shown in figure 8-2.

Antenna Basics

Many strange claims are made about antennas: "The best antenna is one which is a resonant length." Hogwash. "The feed line must be a half wavelength long." Again, hogwash.

The feed line must be long enough to reach from the station to the antenna. That is very basic. In some rare instances, the composite load of the antenna and feed line is such that the Transmatch has trouble achieving a perfect match, or the tuning of the Transmatch becomes critical. In such cases adding about 10 or 20 feet of coiled up open-wire line usually will change the composite load so that the Transmatch can handle it more easily. I suppose this was the original McCoy dipole, but to be honest, amateurs have been using odd-ball dipole lengths for years prior to my calling it the "Real McCoy"!

Here is a big question: "How short can my McCoy dipole be and still be productive?" There is a rule that works pretty well in a general sense for most antennas. The larger (longer or bigger) the antenna, the more gain it will have. Some amateurs use the term *capture area* when referring to the size of an antenna. However, there is no technical basis for this term. An antenna has an *effective area,* sometimes correctly called *effective aperture,* and the larger the effective aperture, the greater the gain of the antenna. I suppose, though, that all this doesn't mean anything when it comes to actual performance.

Dipole Antenna Lengths Versus Performance

From personal experience I have found that one should shoot for something close to one-quarter wavelength (and even less) for the lowest chosen frequency. Example: I put up a half-wavelength long 80 meter dipole, 128 feet overall. I made numerous tests and then reduced the antenna to one-quarter wavelength, or 64 feet. Honestly, the reports on 80 didn't change that much. Because the patterns changed, I was stronger and weaker in different directions. I then went to 32 feet overall (keep in mind that these were all center-fed, open-wire-line antennas). While I noted changes in performance, again they were not what I would call serious. In fact, they were very acceptable.

I then decided to run the antenna lengths through several of my antenna computer programs (MN [MININEC] and ELNEC) using various dipole lengths for comparison. There wasn't a heck of a lot of difference in the quarter-wavelength dipole compared to the half-wavelength. I ran the plots for a 67 foot dipole at 70 feet above average ground and the same for the dipole 33.5 feet long. I noted there is a very slight difference in the patterns. The regular size is tucked in more at the plane of the dipole, and there is slightly less than $1/2$ dB difference from full size to half size. I also ran the same short antenna on 80, but in this case the gain dropped about 5 dB (from front to side), and the vertical radiation pattern changed dramatically with much more high-angle radiation for the short antenna. Nevertheless, and this is the bottom line, even the short antenna worked on 80 meters.

Still another computer run consisted of taking a half-wavelength dipole and dropping the ends $1/8$ wavelength. In other words, I had $1/4$ wave across the top horizontal and $1/8$ wavelength drop. Many, many amateurs encounter this kind of situation, and dropping the ends permits more antenna length. What do we call it? A Drooping McCoy! In any event, the computer showed this to be within a fraction of a decibel of the fully horizontal dipole—in fact, so little

difference it isn't worth running a plot in this book.

My conclusion: Make your multiband McCoy Dipole as long as you possibly can and try to stay away from very, very short lengths. Also, the antenna can be in an inverted-V or sloper configuration. With some of the short lengths I tried, I did find that the Transmatch was more critical to tune, but in all cases I could get a match. To be fair, I even tested the "current" type balanced feed, but surprisingly, the beads got rather hot on the coax when running near a kilowatt (which was with some short antennas and loads). On the other hand, my conventional 4 to 1 transformer/balun using three T-200 cores and Teflon wiring stayed nice and cool.

It is important to understand that the impedances of short dipoles are much lower than those of full-size antennas. Simply, this increases the ratio of power losses from ohmic to radiation resistance losses. However, these ratios are not really significant unless the dipoles are extremely short. Using the open-wire feeders I specified will not change these ratios. For all practical purposes, though, it provides an essentially lossless feed system and efficient operation.

Whether or not the antenna is called a McCoy Dipole, it is still the lowest cost and finest multiband skywire you can use. You only have the cost of the wire and feed line. You can even make your own insulators. It is mighty tough to beat as a performer, and you'll be using a Real McCoy—hi!

To be honest, there are certain antenna lengths that should not be called McCoy Dipoles. A dipole that happens to be the length derived from the formula 468 divided by the frequency in MHz is usually referred to as a "resonant" antenna. One of the reasons I do not put any emphasis on resonant dipoles is simply because we make one for, say, 3800 kHz, and the very first time we QSY away from 3800, the dipole is no longer resonant. It becomes a McCoy, if you will.

Antenna System Resonance

I probably will repeat the following several times, but it is important and must be learned. While the antenna itself doesn't have to be resonant, the *antenna system* must be resonant. Again, resonance is the absence of reactance. Therefore, when we adjust whatever the transmitter is looking at as a load, that load must not have reactance present. In other words, it should resonate. We can achieve this using almost any type of antenna and feed line, but in the antenna system we must always deal with the antenna length as an effi-

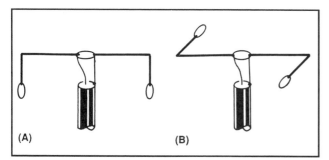

Figure 8-3. Some methods of maintaining "as long as possible" in limited space. This means dropping or bringing back the ends. Amateurs have been doing this for years; it helps. From a technically sound standpoint, however, don't bring the antenna wires back parallel to each other without keeping the wires separated by at least a tenth of a wavelength. Closer spacing can cause signal cancellation.

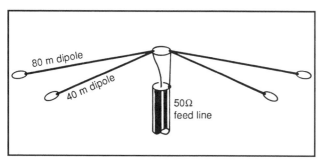

Figure 8-4. The multi-dipole concept with each dipole a half wavelength long, all commonly fed.

ciency factor and the type and losses of the feed line. These are the important points to consider and the simple rules to follow. Coax can be classed as a lossy line if it is not matched to a very low, say less that 3 to 1, SWR. However, even here generalizations cannot be made simply because coax can be low loss and tolerate higher SWRs at low frequencies, while it cannot do the same at, say, 15 or 10 meters and up through VHF and UHF.

The Drooping McCoy

Multiband dipoles can be quite short and still do a fairly good job. The development of the concept of a "Drooping McCoy" may be of interest here.

At one time I lived in a house that had an attic about 30 feet long. I figured I could drop the ends, as shown in figure 8-3(A), about 10 feet, giving me a dipole that was approximately 50 feet overall. I fed this with open-wire line to a Transmatch and used the antenna on 160 through 6 meters!

How did the antenna work? At that time I was inter-

ested in contests and lived in Missouri. As I recall, I had the top, or nearly top, score in the ARRL Sweepstakes and also was first or second place in the CQ World-Wide DX Contest for the state. In essence, this was also an "invisible" antenna because none of my neighbors could see my indoor installation.

As I discussed earlier, this was essentially a no-loss system, because even though there was a high SWR, open wire is a very low-loss line, thereby minimizing losses in the feed line. In other words, all my power was going to the antenna to be radiated. By the same token, the antenna can be laid out in the form of a Z as shown in figure 8-3(B) to create a longer overall length. Bandswitching merely consisted of switching bands and then adjusting the Transmatch for a match.

To repeat, all of these systems require a Transmatch or tuner to be effective. In this regard I am not referring to the tuner or Transmatch types built into modern transceivers. The built-in devices have a very limited matching range, and what is required for tuning open-wire-line systems is a wide matching range Transmatch. However, there are multiband coax-fed antennas you can make that fall within the matching ranges of these built-in tuners.

Multiple-Band Dipoles Using Single Coax Feed

One antenna I described many years ago was known as a single coax-fed multiple dipole. As I pointed out earlier, a single dipole antenna has an impedance of anywhere from 30 to 80 ohms or so, depending on its height. Generally, such a dipole can be fed with 50 ohm cable and achieve a low enough SWR to provide a match for most built-in tuners to handle.

The antenna I concocted consisted of several single-band dipoles, one for each desired band, all insulated from each other over the length of the antenna, but fed with a common feed line at the center. In other words, using insulated wire, I cut a dipole for 80 meters and another for 40 meters, and then fed the two with a common coax 50 ohm line. This is shown in figure 8-4. Individual dipoles should be made from insulated wire (they must be insulated from each other along their length, only being connected together at the feed point). A simple method of making these multiband dipoles is to use lengths of 450 ohm insulated open-wire line (see figure 8-5 for two types). A feed line of lightweight RG-8X type cable can be used for power up to 300 watts or so.

Let's go into some simple technical explanations

here, recalling what we stated earlier about antenna impedances. This explanation applies to many single-feed, multiband antennas, including beams, Yagis, trap dipoles, and so on.

Assume we have a resonant 80 meter dipole, so we know that the feed impedance has no reactance. Let's now attach a 40 meter dipole to the same feed point, still using our single feed, as in figure 8-5. When we feed an 80 meter signal to this antenna, the signal is going to "see" a complex feed point. When we transmit on 80, our 80 meter signal reaches this dual antenna and the signal becomes aware of two impedances. One, the 80 meter impedance has no reactance—nothing to stop the signal. However, the 40 meter antenna impedance is highly reactive to the 80 meter signal, so it won't accept the signal. Hence, our 80 meter antenna does the work. When we switch to 40 meters, the 40 meter signal encounters a similar situation in that the 80 meter antenna is a reactive load, while 40 is not, so the 40 meter antenna goes to work. I have over-simplified here because when we hang four or

five antennas together, we get into some rather complex impedances, and *some* of the other antennas will accept some of the power and will radiate. Here again, though, this is not lost power (all it does is alter patterns) because the RF which will radiate from this "other" antenna will help us work someone we normally wouldn't have worked!

There is one small bonus when using a 40 meter dipole in that it is a more or less resonant antenna on 15 meters, so adding a 15 meter dipole is not needed to get coverage on this band. A 40 meter dipole is three half wavelengths on 15, which will present a reasonable impedance—somewhere near 50 ohms—on this band. Almost certainly any transceiver with a built-in coupler will handle the system, and do it easily.

Using two lengths of the 450 ohm line, one taped over the other (a total of four conductors), will provide five-band coverage. This consists of 80, 40, (15), 20, and 10 meter dipoles. Figure 8-5 provides details of this system. I have made several of these systems, including a four-band multiband ground-plane vertical

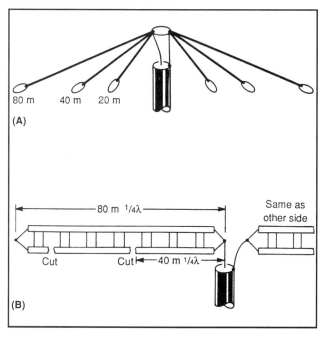

Figure 8-5. Shown at (A) is a system I described back in the early 1950s. It is still an excellent performer. The feed load (transmitter end of the coax) will usually remain at less than an SWR of 3 to 1. This usually can be handled by built-in transceiver Transmatches. At (B) is a method I have used to make multiple dipoles. Using ladder line or 300 ohm twin lead, make a cut in one lead of the ladder line for the correct dipole length. Additional lengths of 450 ohm line cut to proper lengths and taped over the other open-wire antenna will provide coverage for additional bands.

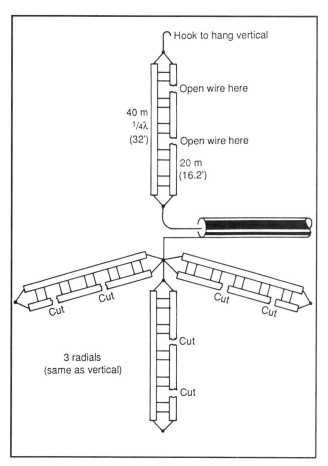

Figure 8-6. Multiband common feed verticals using the same system outlined for the half-wavelength dipoles.

The Uni-Hat multiband vertical antenna. The antenna is 31 feet high with a 15 foot diameter top hat. The antenna covers 160, 80, 40, and 18 meters. Conclusive tests have shown the antenna to be as good as or better than a horizontal full-size dipole.

The Uni-Hat antenna has four sleeve (wire) radiators for 160 meters. These are fed at the bottom into a 50 ohm impedance with a radiation resistance of 45-plus ohms.

(see figure 8-6 for details of this vertical). The three composite radials are also multiband. A 33 foot tall structure (tree, tower, A-frame, end of house with post mounted on it, etc.) is needed to support the antenna. The 33 foot system provides good 40 through 10 meter coverage. Both the horizontal/inverted V and the vertical can be handled by nearly all the "built-in" antenna tuners. I haven't tried the WARC bands, but the SWR is likely to be low enough to include coverage here also. Frankly, from my experiences any of these antennas will outperform trap dipoles. In my many years of dealing with antennas I have never

really liked trap dipoles, but I have known many knowledgeable antenna authorities who do.

The Uni-Hat Vertical

At the time of the writing this book I was asked to review a new antenna—one new in both design and performance. The antenna is called the Uni-Hat, and it is manufactured by the Uni-Hat Corporation. It is a multiband vertical of an extremely unusual and very practical design.

As I discussed earlier, the efficiency of an antenna

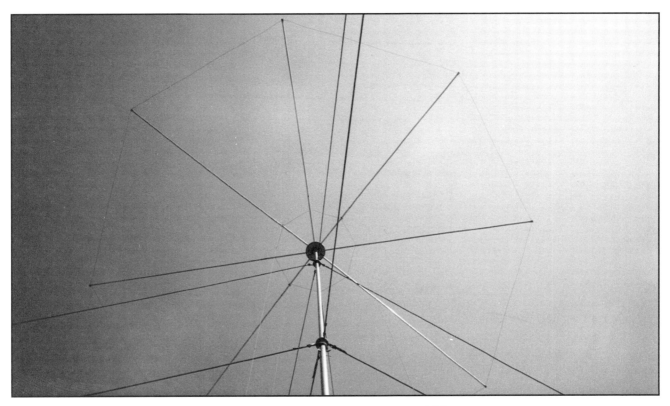

Looking up at the top hat of the Uni-Hat antenna. The sleeve/wire radiators are visible.

depends a great deal on its impedance and the ratio of ohmic losses to radiation resistance. Let me first describe the Uni-Hat and you quickly will see what I mean.

The Uni-Hat is basically a 31 foot tall antenna with a top hat that is 15 feet in diameter. The antenna covers 160, 80, and 40 meters, and 18 MHz under a matched condition of less than 2 to 1. The 2 to 1 bandwidth on 160 meters is approximately 80 kHz, on 80 meters slightly over 100 kHz, and under 2 to 1 for nearly all of 40 meters. The antenna can be used on the other high bands with a Transmatch.

So what is so great about this antenna? After all, there are plenty of trap or loaded verticals on the market. Simply, as I pointed out earlier, if an antenna is shortened physically, the radiation resistance (the useful part of the impedance) becomes so low that the antenna is a relatively poor performer. In the case of the Uni-Hat, the radiation resistance is high—on the order of 50 ohms. Essentially, the antenna is designed to have an impedance of 50 ohms without any matching transformers!

While the design is proprietary, with patent pending, there are a few points I can tell the reader. Essentially, the design uses loop antenna techniques.

On 160 there are four skirt or down wires from the top hat spaced roughly 4 feet apart.

The antenna is fed at the bottom, but here is an unusual point. The radiation or current point is at the top of the four wires. The top hat (see the accompanying photo) is 15 feet in diameter, and the antenna structure uses four sets of guys to support the mast that holds the structure.

As to performance, I have to state that in my many years of testing and using antennas I never had a vertical outperform a good horizontal dipole. However, in this case I consistently got better reports or tied the horizontal in performance on the three top bands. That says a great deal about the Uni-Hat. In fact, in my own case I no longer switch to my horizontal dipole except for tests.

The designers of the Uni-Hat did extensive testing of the angles of radiation from the Uni-Hat on 160 meters. The primary angle is 10 degrees, with a lesser lobe at 45 degrees.

End-Fed Long-Wire Antennas

I have received many requests for information on long-wire antennas. By definition, a long-wire anten-

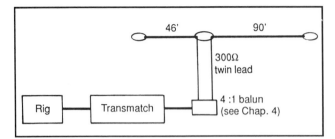

Figure 8-7. The popular off-center-fed dipole. Feed-line radiation can be expected when on 15 meters or some of the WARC bands.

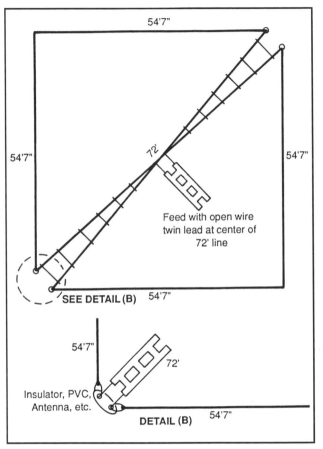

Figure 8-8. The N4PC multiband loop basically is designed for 80 meter low-angle radiation when the antenna is mounted relatively close to the earth. This antenna exhibits good gain in certain directions on the higher bands. It is an antenna I highly recommend if you have the room. (Anyone for 50 foot A-frames?) Detail (B) shows the insulators used in the two corners of the loop.

na is one that is several wavelengths long at the primary, or operating, frequency. Over the years, however, amateurs have come to define a long-wire antenna as one which is end fed, regardless of its length. Keep in mind that any one-quarter wavelength, or odd number of quarter wavelength, wire will provide a low-impedance feed load. We can couple to an odd number of quarter wavelengths much more easily than to an even number of quarter wavelengths. Insulated wire should be used coming into the shack to the rig or Transmatch, and the wire going out of the building must be insulated. The far end of the antenna should be as high as possible—the higher the better. In the case of truly long wires in a straight run at least two or three wavelengths long, there will be a distinct lobe, and gain, in the directions centered around the end of the wire.

The Off-Center-Fed Dipole (Windom)

Many years ago an antenna was invented by a famous amateur, Loren Windom, W8GZ. His design was essentially an 80 meter half-wavelength long wire with a single wire feed about one third from one end. The antenna eventually evolved into one of the same length but fed with 300 ohm line about one-third of the distance from one end. The feed line used is of the ribbon type designed for television use (300 ohm). One slight problem with this antenna occurs on 15 meters, where the impedance becomes rather high. Even though the 15 meter impedance is high, modern couplers will handle the loads. The antenna is very popular probably because the off-center feed is convenient.

Figure 8-7 is a drawing of this antenna. The impedance at the transmitter end of the 300 ohm twin lead is going to be on the order of 300 ohms. A 4 to 1 balancing transformer should get the impedance close enough to 50 ohms so that the average rig will handle the load. There are several very good 4 to 1 heavy-duty transformers available, including one shown with

the description of the Spencer antenna (upcoming). Of course, when using a Transmatch, simply connect the ends of the feeders to the balanced output terminal of the Transmatch and tune it up. The performance of this antenna is just about equal to that of a center-fed, tuned feeder of the same overall length. There will be some feeder radiation (minimal) on 15, but it will be primarily vertical radiation.

The N4PC Multiband Loop

The N4PC multiband antenna requires a little more real estate, but if you have the space I highly recommend it. Earlier I discussed the problem of obtaining a useful low angle from an 80 meter dipole. The 80 meter dipole has to be on the order of 130 to 260 feet high in order to produce useful DX angles. However, a few years back Paul Carr, N4PC, came up with an 80 meter antenna that produced a more useful angle (about 45 degrees) for an 80 meter loop only 30 to 50 feet high.

This antenna is a very good performer on 80, and by using open-wire feeders it can effectively be tuned up on any of the other bands, producing radiation lobes that have good gain. The antenna is shown in figure 8-8. Each side of the antenna is 54 feet 7 inches long, and the opposite corners of the square are open. Connecting across these corners is a 450 ohm insulated open-wire phasing line 72 feet long. Note the line is transposed one-half turn. At the exact center of the phasing line the feed line (same type 450 ohm line) is connected and brought to the station Transmatch.

The loop is more diamond shaped than square because of the phasing line length. This antenna has been thoroughly tested. The overall pattern of radiation in the horizontal plane is basically omni-directional on 80 meters. But where this antenna separates the men from the boys is in the vertical plane. With the antenna installed at 30 to 50 feet above average ground, the primary vertical take-off angle is on the order of 40 to 45 degrees. When used on 40 meters, the take-off angle drops to 35 degrees and the horizontal pattern changes. Of course, these lobes will have some gain. However, neither Paul nor I would want to go out on a limb and put definite gain figures on the antenna. As we go up in frequency, through the higher bands, the take-off angles get lower and lower, with increasing gain.

If you think I am showing a great deal of enthusiasm for this excellent multiband antenna, you are correct. It is terribly difficult to come up with any useful DX antenna angles for 80 and 40 meters simply because for these bands 100 foot or higher towers are required.

The Spencer W4HDX Multiband Dipole

One of the most unusual multiband dipoles, and an antenna I like very much, is the Spencer antenna. This is normally an off-center-fed dipole (common enough). What is unusual is the dipole is fed with RG-11, 70 ohm coax using tuned dual feed lines. It also can be center fed if certain antenna lengths are used.

Normally, I would not recommend tuning a coax line for several reasons. The most important of these reasons is that certain precautions must be observed to avoid a high standing wave ratio on coaxial lines. Keep in mind that the SWR is the ratio of the maxi-

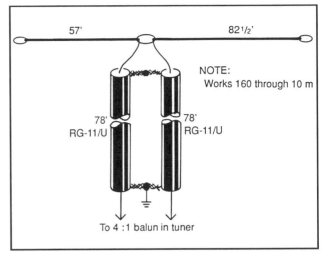

Figure 8-9. The off-center-fed Spencer antenna is shown in this drawing. This feed method is highly recommended for noisy locations, or when balanced feed is desired and the feed line must be run up ducts, close to metal structures, and so on. It is a very good performer.

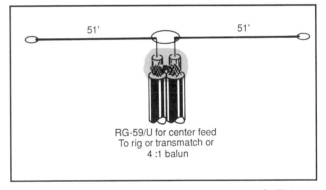

Figure 8-10. The Spencer type of antenna in a G5RV configuration. It is quite likely a Transmatch will be required here.

mum to minimum voltages that appears on a line. If you run high power and the SWR is very high, there is a real danger of blowing out the line or actually melting it. Not only that, but losses increase severely with a high SWR on coax. However, and keep this in mind, if the SWR is less than 5 to 1, either the RG8 or RG11 type coax can easily handle the mismatch.

What is so special about the Spencer antenna? It has several points that make it different from a conventional open-wire tuned line antenna. Figure 8-9 shows the off-center configuration, and figure 8-10 shows a configuration using center feed—the popular G5RV 102 foot dipole. Either of these antennas provides a relatively low feed impedance that will ensure a reasonably low SWR for the dual coax feed. Spencer recommends using dual lengths of RG-11/U (a 70 ohm

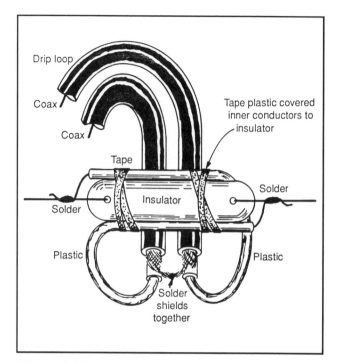

Figure 8-11. Recommended method of supporting dual coax lines. I would also recommend using sealant material on the ends of the shields to prevent moisture from getting into the lines. For power up to 300 watts, it would appear that the smaller RG-8X type of coax would be suitable. For power up to the legal limit, I would recommend a very good grade of 50 ohm, higher voltage cable.

impedance coax) in the off-center configuration. It is permissible to use 50 or 70 ohm coax for the center-fed 102 foot dipole, but in all likelihood a Transmatch will be required. In this case, for powers up to 150 watts or so the popular RG-8X types could be used. I would not recommend RG-58, as it might get a little dicey if you encounter a high SWR condition.

The off-center-fed antenna is recommended for rigs with built-in Transmatches. The antenna is 57 feet on one side and 82.5 feet on the other, fed with dual 70 ohm RG-11 coax. The dual coax is connected to a 4 to 1 balancing transformer. From the transformer, RG-8 is connected to the rig. The SWR is below 3 to 1 on all bands—including the WARC bands. This is at the transmitter side of the balancing transformer. In this case it is not likely that an external Transmatch would be necessary. However, we have no way of predicting what an individual's ground conditions are or what the height of the antenna will be. As pointed out earlier, the impedance of an antenna is determined by many factors, and the antenna height is one of the most important considerations.

One of the problems with built-in tuners is that they

have a rather limited matching range; not much more than a 2 to 1 mismatch can be handled. Therefore, because of antenna heights and impedance changes, a wide-range Transmatch such as those described in the Transmatch chapter may be needed.

Getting back to the Spencer, as can be seen from the drawing, only the *inner* conductors are used to feed the two halves of the antenna. The outer shields of the coax are connected together at the antenna end but not connected to anything else. (See figure 8-11 for an appropriate method for supporting coax cables at the antenna.) Back at the Transmatch end, the shields are also connected together and then grounded to the Transmatch. Therefore, only the inner conductors of the coax are used to carry RF. Most important in this antenna, these two inner conductors are shielded all the way from the antenna to the station. What is the advantage of using this coax over using open-wire feeders?

Consider the following. Suppose we are living in a condominium or apartment. We want to put an antenna on the roof, but the only way to get the feed line up to the roof is via a pipe or elevator shaft. Or maybe we wish to "hide" the feeders by running them alongside the rain gutter or downspout. Here is a *multiband* antenna that gives the perfect answer. The feed line can be snaked alongside any metal object, such as a tower, or in a ventilation shaft, etc., and we will not need not worry about the feeders being affected. Also, several types of coax can be buried, which provides additional flexibility and other possibilities.

Another advantage of this antenna and feed style is its capability of handling noise problems. In a single-coax-fed dipole with the inner conductor going to one side of the dipole and the shield to the other, any noise picked up on the shield can be coupled to the center conductor and back into the station receiver. When Spencer went to the dual-feed system, the noise practically disappeared. What is meant by "practically"? Spencer states that if there was an S8 noise level with a single coax, use of the dual-feed coax would drop the noise to S2 or less! This in itself is an extremely worthwhile feature. I personally had a bad noise problem at one time (power line), so I tried the system. It knocked down the noise as much as four S units!

The dual-coax shields are grounded to the Transmatch or to earth ground, and the two inner conductors of the dual coax are connected to the balanced line input. The Transmatch is then adjusted for a match. In some odd-ball feed-line lengths it may be impossible to achieve a perfect match. Anything less than 2 to 1 is suitable, but by adding some feed-line length it is possible to change the matching load and get a perfect match. Spencer has said that the off-center version does a fair job on 160 and exhibits excellent performance from 80 through 10, including the WARC bands.

I have tested the Spencer antenna thoroughly, and I can verify that it is a good performer, particularly in a noisy location. Would I recommend it over a multiband dipole fed with open-wire line? The answer is yes and no. The Spencer antenna has several features that open-wire line does not. The fact that the line can be buried, taped to a metal tower, etc., is in its favor. It is much more expensive than the coated open-wire/ladder line, and of course, open-wire line has no real measurable losses regardless of the SWR, However, the Spencer does provide answers for tough installations. On the other hand, open-wire line with a wide-range Transmatch will handle *any* SWR or mismatch without noticeable loss, and as I have pointed out, the antenna can be *any* length. However, I don't recommend trying to run open-wire/ladder line inside metal pipes or on tower legs in the same way that feed lines can be installed with the Spencer. See chapter 4 for suggestions for a 4 to 1 balun for either the Transmatch or "direct" feed.

Multiband Verticals

A subject with which I have not dealt in great detail is that of multiband vertical antennas for fixed locations.

A little later in this book I treat mobile verticals, but here I would like to cover multiband verticals of a commercial nature that could be used as the "home" antenna.

Vertical antennas have always been popular simply because they are generally inconspicuous and take up very limited real estate. I have the good fortune of being familiar with nearly all the commercial units. Here again, without being too critical, you should be aware of some of the problems involved in trying to make an "all" band antenna. These include problems such as traps have losses, reduced physical size means reduced gain, and so on. All explanations of how trap beams and dipoles work apply equally to trap verticals.

Probably the most popular commercial trap vertical antennas are the Cushcraft R5 and R7. I tested both of these and found them more than adequate. In fact, in testing the R7 I managed to work over 100 countries with 100 watts on CW during the CQ World-Wide Contest. True, my signal wasn't a world beater, but it did the job.

Another popular line of verticals is made by GAP Antenna Products. These antennas use a concept different from "normal" verticals in that they are fed at the center of the vertical rather than at the base. The latest model GAP antennas have their own "counterpoise," so extensive radials are not required. I found these antennas to be excellent radiators.

Another antenna I like is the Butternut multiband vertical, which also does an excellent job.

Chapter 9

Multiband Rotatable Beams

I am not going to attempt to show you how to build a trap beam. There have been many articles on the subject, but none that I know of in recent years. The very first article was by Buchanan, W3DZZ. He described a three-band trap beam (plus a trap multiband dipole) in a June 1955 article in *QST*. In my mind, this was a case of the first being one of the best. If you want to "roll your own," I would strongly suggest a trip to the library or a visit to someone who has that issue of *QST*.

A Short History of Trap Beams

I would not want to create the impression that trap beams, including Yagis, are not good antennas, but there are facts that everyone should know.

I was present when the first trap beam came into existence. Shortly after Buchanan's article appeared, commercial manufacturers started making and marketing trap antennas. There are not many of us around anymore from that period (some 40 years ago)!

In years past I had the opportunity to measure some trap beams at reliable antenna ranges (with excellent equipment). Many of them showed a 6 dB forward gain on 20 meters and slightly more gain on the higher bands. Some were pretty bad, though. In fact, one or two of the 1950s era had no gain at all. They actually had minus gain figures! What you should under-

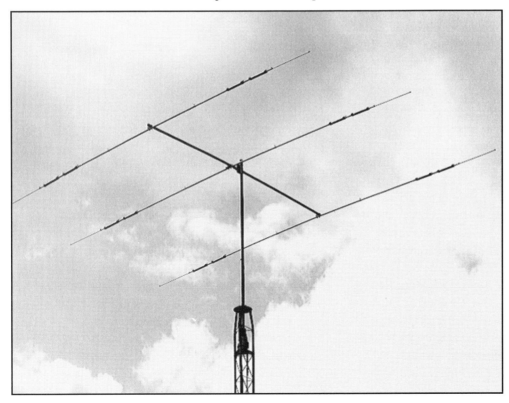

This is the Cushcraft three-element 20, 15, and 10 meter trap beam. I used this antenna for several years with great satisfaction and would not hesitate to recommend it.

A three-element monoband beam for 10 meters. This Cushcraft monobander will produce a 7-plus dB signal and give a very good account of itself.

stand is antenna engineering includes a lot of black magic. The problem then is the interpretation of performance figures for presentation to potential users. This can be a very creative process.

For the sake of clarity, let's look at gain and how important these gain claims are in the real world.

Let's assume a three-element monobander has 7 dB gain. A good trap beam has 6 dB gain. An S unit on a calibrated receiver is usually considered equal to 6 dB. The difference between 6 and 7 dB normally isn't worth considering. You are not going to lose out in a pile-up with those differences, assuming you have a good trap beam.

Before I forget, there is something of which you should be aware: Doubling the size of a monobander, as I already have pointed out, will give 3 dB more gain. However, when doubling the size of a trap beam, the same formula does not hold. The larger trap beams will still show a gain around 6 dB, because the complexities of element spacing, trap losses, etc., preclude an easy way to obtain a reasonable impedance. You might gain a fraction of a single decibel, but that would be all. I have seen antenna-range results where a six-element trap beam had only a fraction of a decibel more gain than a three-element trap beam!

How do you become an astute amateur and know which trap beam to buy? One simple method is to find out what successful stations are using. What is a successful station? Obviously, anyone who consistently is a good performer in contests fits into this category. Look up the calls in the *Callbook* and write to them or call them.

Design Problems of Trap Beams

Getting back to our discussion of trap beams, let me state some of the problems involved in designing them. From a design standpoint, the main criteria at which the beam manufacturer looks are bandwidth and making an antenna that always will present a 50 ohm load. Unfortunately, it is an impossible project to design a multiband trap beam that will present an SWR of less than 2 to 1 on all frequencies the beam covers. Remember what was said earlier: The modern transceiver manufacturers design their equipment to work into loads of 2 to 1 SWR or less. If the load exceeds that, the transmitter shuts down. This leaves the antenna designer with a requirement to design an antenna that meets the transceiver load specs—an impossible task for several reasons.

First, for traps to be effective, and actually divorce one band from another, they must be very high-Q devices. What is meant by "high Q"? In this case, I am talking about a device that is very selective.

Suppose we are transmitting a 15 meter signal into a trap beam tribander covering 10, 15, and 20 meters. We want our 15 meter signal to "see" only a 15 meter beam—electrically, that is. When the 15 meter signal enters the driven element, it first sees the 10 meter trap. But because the 10 meter trap has low reactance to a 15 meter RF signal, the 15 meter energy acts as if there were no 10 meter trap present. Now when the 15 meter signal reaches the 15 meter trap, that trap acts as a complete blocking device which prevents the signal from going farther out on the element. Therefore, the trap is designed to be very selective. This selectivity is referred to as high Q. (A low-Q trap would be very broad.) Keep in mind that for the triband beam to act like a 15 meter beam, the RF must be kept out of the 20 meter portions of the antenna. Why? Simply, in this case a 20 meter beam fed with 15 meter energy would have little or no gain on 15, and the front to back, front to side, etc., would be destroyed. Therefore, if the traps are not designed correctly or are not manufactured properly, they will introduce losses and decrease the desired bandwidth.

From yet another standpoint, a monoband three-element beam will have a relatively low impedance. For example, a close-spaced beam with one-tenth wavelength element spacing will have an impedance of just a few ohms. A monoband beam will require a matching device to get this low impedance up to 50 ohms to match the coax.

In the case of a multiband trap beam we must consider complex impedances formed by trying to use more than one band. We can increase these impedances on a trap beam by juggling element spacing and element lengths. *However,* we no longer think in terms of gain. We merely think in terms of matching to provide a flat load!

Another problem involved is the antenna is designed so that its feed impedance, with matching network, will work at a certain height above ground. The height, as pointed out earlier, is a major factor in determining impedance. The manufacturer may have made the assumption that the buyer is going to put the beam on a tower at 50 or 60 feet, but the manufacturer has no way of really knowing how high an installation will be made. The same beam checked out by the manufacturer at 50 feet and 50 ohms impedance is going to be a lot different in the buyer's location, say at 30 feet. In addition, structures located near the installation site of the antenna may affect performance.

As an aside here, many amateurs have asked me if I use a Transmatch with beams even though the SWR

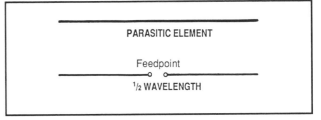

Figure 9-1. The original design of the Yagi consisted of a dipole driven element, with a parasitically excited dipole to the rear.

doesn't go above 2 to 1. The answer is definitely *yes.* I always want my transceiver working into its design load of 50 ohms. Plus, my receiver works better with the Transmatch in the line, because the receiver is also designed for 50 ohm input.

Yagi Antennas

The idea of a directional beam antenna using parasitically excited elements was first discussed by Yagi and Uda, two Japanese scientists. (There were directional multi-element beams before this time, but it was customary for all elements to be "driven," or directly connected from the generator or transmitter.)

Yagi and Uda set down the concept of using more than one half-wavelength antenna element to obtain directional patterns. One antenna element was "driven" by attaching the feed line from the transmitter. A second element or dipole spaced a specific distance behind the driven element was parasitically excited by the field from the driven element (see figure 9-1). Power was fed with a direct connection to what we now define as the "driven" element. It was found that by using two elements, one driven and the other parasitically excited, a 4 to 5 dB gain could be achieved in the desired direction as compared to a single dipole. In addition, signals coming from several directions could be attenuated, giving us what we call "front-to-back and front-to-side" signal attenuation.

In comparing gains, comparisons of a real antenna to an isotropic antenna provide bigger numbers. A three-element monoband Yagi would have on the order of 7 dB gain as compared to a dipole "dBd" (note that the last "d" in dBd means dipole). If we compare the Yagi to the isotropic, we would add 2.14 dB and come up with 9.14 dBi, the "i" meaning isotropic. Amateurs like to get the most gain for their money, so it is wise to know what is being compared to what. In this book all the gain figures mentioned are in dBd (dipole comparisons).

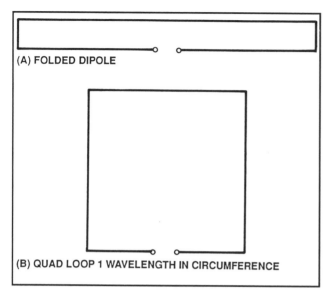

(A) FOLDED DIPOLE

(B) QUAD LOOP 1 WAVELENGTH IN CIRCUMFERENCE

Figure 9-2. When Moore designed the first quad, he went on the assumption of opening up a folded dipole. At (A) is the folded dipole. The total distance around a folded dipole is one wavelength. In the quad the same length of wire is used, but it is made in the form of a square (or diamond) shape. Normally, we think of current point in an antenna being the area of maximum radiation. In the antenna shown at (B) the current points would be at the feed point and at the top center. Since the sides of the loop are one-quarter wavelength long, in effect we have stacked the two current points one-quarter wavelength apart. Antenna, or dipole, stacking increases gain and changes the effective angles of maximum radiation.

Getting back to our discussion of beams, as more parasitic elements were gradually added to the Yagi concept, the gain of the beam increased. Usually there was one slightly longer element used as a reflector, and slightly shorter elements were used as directors. Again, these added elements are all referred to as "parasitically excited." It was found that a driven element used with a reflector and director would shape the signal into a directional pattern that had approximately 7 dB of gain compared to a single driven element or single resonant dipole. The driven, reflector, and director elements were all on the order of one-half wavelength long with the reflector being slightly longer and the director slightly shorter than the driven element.

(One other important point for newcomers to consider when discussing beams is that a three-element monoband beam will produce about 7 dB of gain.)

This original Yagi-Uda design evolved into the Yagi-type beam, which usually consists of three elements with a gain on the order of slightly more than 7 dB compared to a dipole. This is for a single-band beam, not the popular multiple-band trap beam. However, there are other beams which must be considered in order to more fully cover this topic.

Multiband Quads

One type of beam I have always liked is the multiband quad. Let's look at the history of quads.

In this case again, I have personal knowledge of the time period during which the quad was created and of the people involved in its creation. Just before World War II a missionary radio station was built in Quito, Ecuador. The primary designer was Clarence Moore, W9LZF. The altitude of the station was over 10,000 feet, a serious factor when using a Yagi-type beam. The Yagi-Uda designs are high-Q antennas, which without getting too technical, have a bad habit of developing corona discharge from the element ends at high altitudes. Moore and his people designed and built a large, full-size, three-element Yagi for approximately 25 meters. When they applied power to it— 10,000 watts—corona arcs caused the antenna to burn up and destroy itself. Moore tried several methods to reduce the voltage coupling from the element ends. He even tried using copper toilet flush floats on the elements. Nothing worked, however, and the antennas continued to self-destruct.

Moore rationalized, and very correctly I might add, that an antenna with "low Q" elements would not have this problem. He therefore constructed a beam using full-wavelength elements consisting of closed full-wave loops. It took considerable research, but he came up with a full-wave loop in a diamond configuration, fed at the bottom, closed at the top. He told me at one time that he actually conceived of the idea of quads by considering the feasibility of "opening" up a folded dipole. I have shown the basic quad configuration in figure 9-2. Moore reasoned that this would be a low-Q antenna, and quite likely would not suffer from the high-altitude corona problems that the Yagi did. He was right. More important, he created a new antenna design, the quad, which was very effective in performance. Despite its excellent performance, the quad was, and continues to be, the basis for technical disagreement and arguments to this day.

After creating the basic concept, Moore then added elements for a more directional antenna. His measurements disclosed that by simply using a driven element (full wave in size) and a single reflector (full wave), he could very closely approach the 7 dB gain of the three-element, half-wave element length Yagi.

Moore found several features in this new antenna which made it a worthy competitor of a Yagi. Rain and snow striking a Yagi in many instances created very heavy static simply because of the electrical charge in snow or rain. This did not occur in the quad. It is worth mentioning here that there were several quads used by the military in the Desert Storm operation. Sand storms, with their charged sand particles, made the use of Yagis practically impossible, while with the quads the noise was not present. The basic two-element quad is also usually much lighter in weight.

Moore had a very difficult time convincing the antenna engineering society that he had made a worthwhile contribution to the art. I have always felt sorry that Moore never received the recognition he so richly deserved. The quad did become very popular with foreign stations, however, simply because the construction materials were easy to obtain and the antenna was inexpensive to make.

As an aside, I knew Moore personally. We were not exactly next-door neighbors, but he lived in Indiana only about 100 miles from me when I was W9FHZ living in Illinois. At that time a large group of DXers hung out on 28.500 MHz (10 meters) chasing DX, and Moore, when he was home from Ecuador, was one of them. Obviously, there was a lot of competition and serious discussion about quads versus Yagis long before the quad really became popular. I belonged to the "wide-spaced" Yagi school because several of us had discovered that wide-spaced element Yagis performed better than close-spaced ones. I will admit that then many of us did not know a lot about antennas, and certainly very little about beams. This was really the beginning of the directional beam era. However, Clarence Moore with his two-element quad took DX away from many of us even when we were using four-element beams. I also might add that we all were running the legal limit, which was one kilowatt input to the final amplifier in those days.

It was only a few years later that I went to work for the ARRL in the Technical Department. I found that although the department had checked out a single quad loop and found that the loop had nearly double the decibel gain of a dipole, the ARRL was not enthusiastic about a quad. The actual gain of a single full-wave loop is 1.8 dB over a half-wave dipole, but I must admit it was a cumbersome antenna to build, and in those days it was not really strong structurally.

While the following has nothing to do with a technical discussion of quads and Yagis, I cannot help but mention it. When I moved to Connecticut and went to

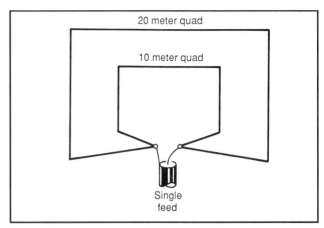

Figure 9-3. Two quad elements (20 and 10 meters) fed together may or may not cause a change in actual radiation patterns because of power division in the loops. When on 20 meters, the 10 meter loop presents a "reasonable" reactance. This tends to divorce the 10 meter loop from the 20 meter loop. On 10 meters the 20 meter loop has an impedance that is higher in value. However, the 20 meter loop will still radiate 10 meter energy. A two-wavelength loop has a different pattern than a single-wavelength loop. There is certain to be some interaction. As pointed out in the text, whether this is bad or good is at present anyone's guess. To remove doubt, the simple answer may be feed-line switching.

work at the ARRL, I was not making a lot of money. I was married with two daughters, and so like many, I didn't have much money for towers, antennas, and so on. At the time I discovered that Persian rugs were shipped from the Middle East wrapped around long bamboo poles. It so happened that there was a rug store just across from ARRL headquarters. I made friends with the store owners and soon had a ready supply of these poles, which made excellent supports for the wires of quads. Believe me, I made quads (plus scores of other weird antennas)! I honestly believe that at the time I probably garnered more experience with quads than most amateurs. However, while all this is somewhat related, I am getting away from the real purpose of this book.

I have long argued that for a given boom height a quad has a "slightly" lower angle of radiation than a Yagi. There is a current stacking effect in a quad element based on effectively having two half-wave elements with the current points at the bottom and top. Such stacking produces a little more power at a slightly lower angle. I know there have been tests made (over short periods) that argue I am mistaken. However, knowledgeable DXers and most old timers will tell you that because of a slightly lower angle of radi-

Here is a five-band, two-element quad made by Antenna Mart. I use this antenna and find that it is a fine performer.

Another view of the Antenna Mart five-bander.

ation, the quad will "open" a band to DX earlier and keep the band open slightly longer than a Yagi at the same boom height. I agree with these people, and I have a very good reason for doing so.

For a number of years while I was employed at the ARRL I was also engaged in a program of "monitoring" several foreign broadcast stations. I used a two-element quad at a boom height of 50 feet. At the same time another amateur who also worked at the ARRL was engaged in the same program, monitoring the same stations. He used a Yagi at exactly the same boom height—50 feet. Without exception, I could hear the stations earlier and later than the other amateur could. This was a test conducted over a two year period consisting of well over several thousand stations being monitored! I'll be the first to admit that this was empirical testing, but it sure was one very long and conclusive empirical test.

Since those days a lot of work has been done on quads and a great deal has been learned. For example, the gain of a two-element quad, driven element and reflector, is slightly less than 7 dB—about the same as that of a three-element Yagi. Also, because of the low

Q elements, the spacing of the quad elements is not as critical as the spacing of Yagi elements.

While the next point has been argued frequently over the years, I personally stick by my guns: For the same boom height, the quad has a lower angle of radiation than the Yagi—on the order of a few degrees. I must be very honest, however, and state that as to actual gain, there was very little difference in received signal strengths between my quad (multiband) and the three-element Yagi. However, note that I said my quad was a three-bander—20, 15, and 10 meters.

What about multibanding a quad? It is possible to go to all five amateur bands (20 through 10) on a single structure. This can mean five driven elements, all fed together with a single length of 50 ohm coax, and five separate reflectors.

I think other writers have avoided talking about this when discussing quads. For example, take the case of

If you want to be a Big Gun, then this is the antenna that will really do a job for you. It is Antenna Mart's five-element, five-band quad. It is an outstanding performer on the 20 through 10 meter bands. It has lots of gain and lots of versatility. This antenna uses individual feed for each band, but only a single feed to the tower. A high-power switch, mounted at the antenna but controlled from the station, is used to switch feed to individual beams.

a 10 meter and a 20 meter quad fed together with a single feed line (see figure 9-3). Keep in mind that a quad driven element is a full-wave loop with an impedance of 100 ohms or so. The 10 meter quad loop (100 ohms impedance) is a more or less resonant antenna fed with 50 ohm coax. Mounting a 20 meter full-wave quad loop around the 10 meter loop adds a two-wavelength loop to the 10 meter loop. The 20 meter loop will have an impedance of about 250 ohms when fed with 10 meter energy. Assuming transmission on 10 meters, will the 20 meter loop also accept power? Simple Ohm's law tells us that while 10 meters is the lower value, 100 ohms, power will also be applied to the 250 ohm 20 meter impedance. With two more or less resonant antennas tied to the same feed line there is no doubt that the 10 meter power will go to both loops. How much power and how much the pattern is changed are very good questions. Frankly, I don't have the answer, and I don't think anyone does. It would take some fancy computer modeling, and we haven't even started talking about the complications introduced by adding a reflector or director to this quad.

The Five-Band Quad

Let's jump to a quad I reviewed in *CQ* ("*CQ* Reviews: The Lightning Bolt Quad," April 1993). This is a five-band antenna with driven elements and reflectors. It uses proportional spacing to obtain the best gain per

band. However, five driven elements are connected to a common feed point. I am guessing here, but I assume that the total impedance of the array comes out to about 20 to 25 ohms. (For the benefit of any antenna engineers reading this, I am not really overlooking the various reactances lumped together here). After all, we are connecting several (five) 100 ohm impedances in parallel, and Ohm's law applies.

The beam in question makes use of a Jerry Sevick 2 to 1 type transformer that works almost perfectly to get from 25 to 50 ohms with no problem. I very carefully checked this beam for impedance matching on all five bands, and the worst case was only slightly more than 1.5 to 1 SWR! However, I personally still prefer to switch the feed line to each beam to provide some isolation and to make sure that the loops are not interacting.

Another five-bander, using separate feed lines, is the Antenna Mart unit shown in the accompanying photo. Both the Lightning Bolt Co. and Antenna Mart sell the parts for complete antennas. Both antennas are very good. I must add that as far as meeting the 50 ohm load requirement, these quads are very flat—less than 2 to 1—on all bands.

Quad Patterns

All of this still does not answer the question of patterns. What I can say from tests on this antenna and many, many other multiband quads using single feed is that there doesn't appear to be too much or too serious a pattern disruption. There is some pattern degradation of the 10 meter beam in a 20/10 combination antenna, but not to an extent I consider serious. Using a computer program on the 20/10 combination, there appears to be a slight clover-leaf pattern on 10 meters. In actual practice, however, it doesn't show up—at least not to a point where it can be checked on either ground wave or DX.

Computer Modeling of Antennas

More and more amateurs are depending on computer-derived information for antenna construction. Early on I wasn't happy with what I was seeing from some of these programs as far as quads were concerned. I now see that others agree with me. One of the finest antenna programs is ELNEC, by Roy Lewallen, W7EL. I would like to quote from the documentation in Roy's program.

"Experiments indicate that quads require a large number of segments (perhaps 12 or more per loop side) to give a reasonable representation of front-to-back ratio. When dealing with wires connected at an angle, MININEC (and ELNEC) 'cuts the corner' by half a segment length. This doesn't cause much error with single loops but apparently causes enough change in the relative currents in multiple loops to quite noticeably affect indicated front-to-back ratios. The forward pattern and gain are *fairly [Italics are mine.—W1ICP]* accurate, with 6 segments/side or so. The segment-tapering method is an effective but rather tedious way of improving the accuracy of front-to-back indications. I am investigating methods to improve quad modeling but haven't made any breakthroughs yet."

I am sure that if anyone can make "breakthroughs" it will be Roy. The business of quad modeling—and statements you hear from amateurs who "computer model" quads—should, in my opinion, be viewed with care.

Radiation Performance of Two-Element Quads

Getting back to the technical discussion of quads, the impedance of a two-element quad is approximately 100 ohms. A 2 to 1 (100 to 50 ohms) Sevick-type transformer/balun brings the match down to 1 to 1.

For the single-feed-line, multiple-element quad the bottom line is there has to be some power radiated by

both antennas at the same time. I could easily pose some "cute" guesses here, and I wouldn't mind hearing from quad people as to what your "guesses" would be. For example, assume a single feed line to a 20/10 quad being used on 10. Would the gain be more or less than that of a 10 meter quad by itself? Keep in mind that the 20 meter part of the antenna is a larger antenna on 10 with a more effective aperture. I could go on, but I would prefer you open your mind and be inquiring. Of course, one answer is to use a single feed line with an antenna feed-line switch such as those manufactured by Ameritron or Antenna Mart (that is what I do). This would effectively change the antenna to a single feed antenna instead of a common feed type.

When discussing common feed, what about losses caused by the quads that are not in play? There are bound to be some ohmic losses. By the same token, though, because the loops are made from good low-loss wire, the losses cannot be great. In fact, they are almost not worth mentioning. So on the basis of losses, when comparing a multiband trap Yagi to a quad, apparently the quad will come out the winner.

Decision Factors—Quad Versus Yagi

One of the negative statements about a quad that is no longer true is it is not a good antenna for high winds or icing. That was true in the days when quads were built with bamboo poles, because the bamboo did not hold together in storms. However, quads built with fiberglass supports are every bit as strong as aluminum Yagis.

There are a number of factors that enter into the decision to select a quad or a Yagi. The first is weight. A quad will come in at a much lighter weight than a Yagi. This means a lighter weight rotator (lower cost) will handle the quad. Point two is a five-band quad is much less expensive than a five-band Yagi, likely one third the cost. Also, the turning radius of a two-element quad is much less than that of a three-element Yagi. Last, but by no means least, the gain for each band will be the same—slightly less than 7 dB. Naturally, all of these assessments are my own opinions based on many, many years of hands-on experience. I feel that in all honesty I can determine the good ones from the bad ones.

Quad Tuning

Those amateurs who have had the most success using quads are those who have tuned the quads—usually just the reflectors—either for maximum front-to-back or for forward gain. This requires field-strength meters and other special equipment to do the job right. One of the most difficult requirements is installing the quad so that you can reach the reflectors for tuning. This even may require hiring a power-company lift chair! If you do attempt to tune a quad, set up a dipole at least five wavelengths away with a simple field-strength indicator on the dipole.

Quad Materials

What kind of wire should be used? I would suggest buying wire from either Lightning Bolt quads, The Wireman, or Antenna Mart. These companies are very reliable and have been making and selling quad parts, including wire, for years.

There are some construction tips worth knowing. From years of experience, it appears to be a good idea to "lock" the wires in place where they go through the fiberglass support rods. This technique is shown in figure 9-4. Using the twist of extra wire tends to lock the quad element wire in place, preventing flexing and eventual breaking.

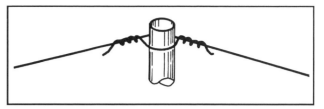

Figure 9-4. This shows how quad wires are "locked" in place. There are two schools of thought as to what the best methods are for keeping wires from flexing and breaking. In my experience the method shown here has proven effective.

Quads on The High Bands

Quads are also excellent antennas for VHF and UHF applications. When using quads on these high bands, you must decide whether you want vertical or horizontal radiation. For vertical radiation the feed line enters at one side of the driven element, not on the bottom. By doing this the quad becomes vertically polarized and can be used for repeater or packet operation. Normally, on the lower bands (20 through 10)

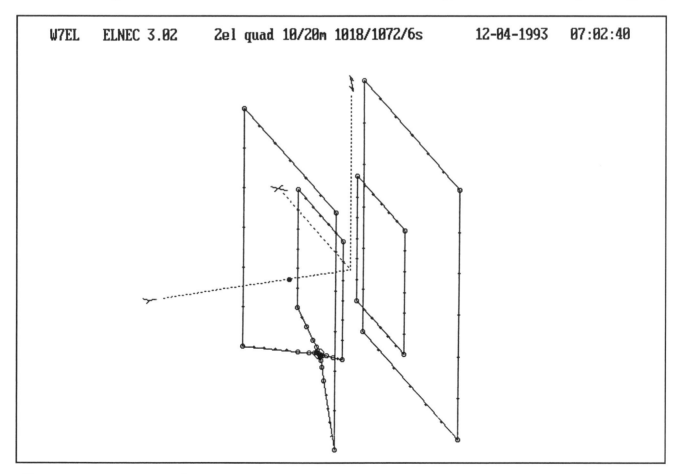

Figure 9-5. This is another configuration of the 20 and 10 meter quads discussed in the text, with common feed.

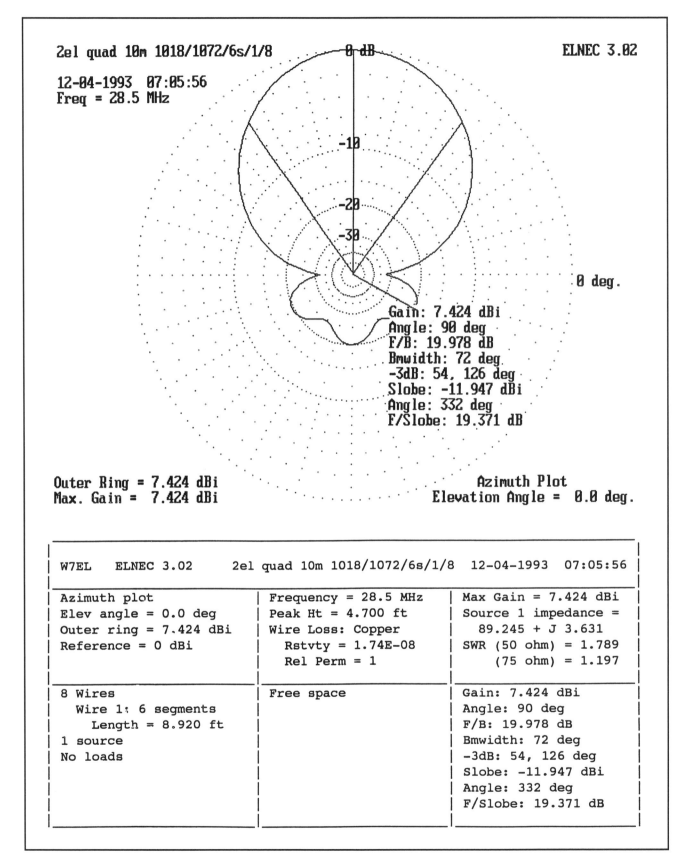

W7EL   ELNEC 3.02   2el quad 10m 1018/1072/6s/1/8 12-04-1993 07:05:56

Azimuth plot	Frequency = 28.5 MHz	Max Gain = 7.424 dBi
Elev angle = 0.0 deg	Peak Ht = 4.700 ft	Source 1 impedance =
Outer ring = 7.424 dBi	Wire Loss: Copper	89.245 + J 3.631
Reference = 0 dBi	Rstvty = 1.74E-08	SWR (50 ohm) = 1.789
	Rel Perm = 1	(75 ohm) = 1.197
8 Wires	Free space	Gain: 7.424 dBi
Wire 1: 6 segments		Angle: 90 deg
Length = 8.920 ft		F/B: 19.978 dB
1 source		Bmwidth: 72 deg
No loads		-3dB: 54, 126 deg
		Slobe: -11.947 dBi
		Angle: 332 deg
		F/Slobe: 19.371 dB

Figure 9-6. This is a pattern derived using a single 10 meter quad, two elements, fed by itself. This should be compared with figure 9-9.

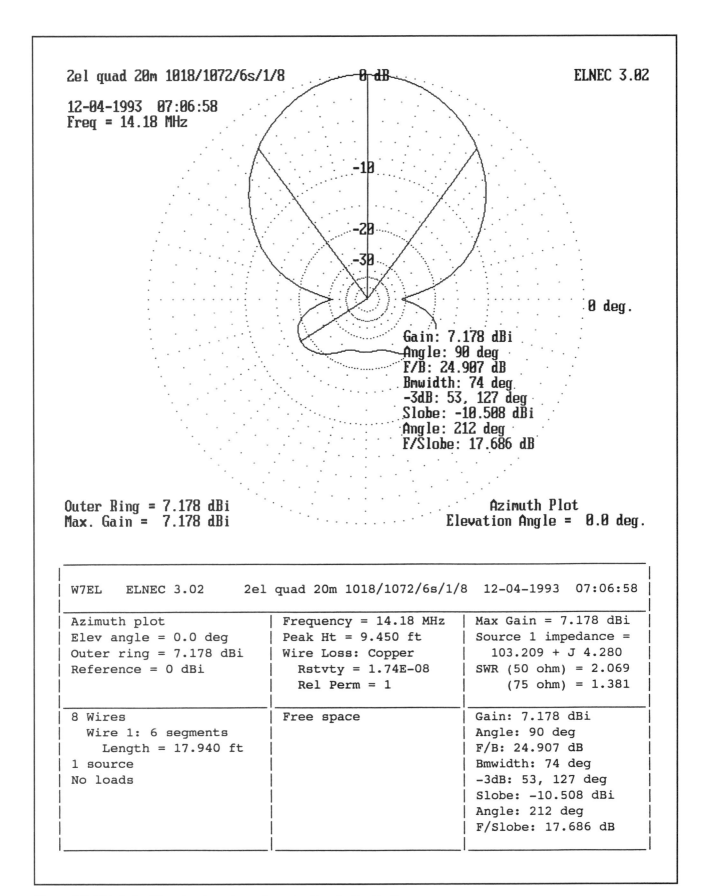

Figure 9-7. This is the 20 meter, two-element quad pattern fed by itself.

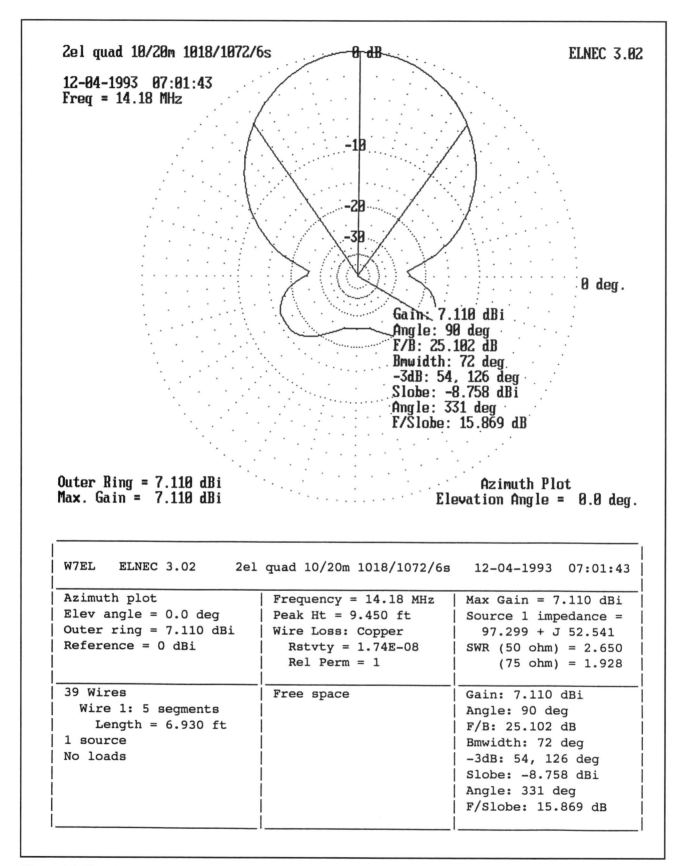

Figure 9-8. Here is the 20 meter pattern when fed in conjunction with a 10 meter quad. Note the very slight difference in gain and also slight differences in front to side.

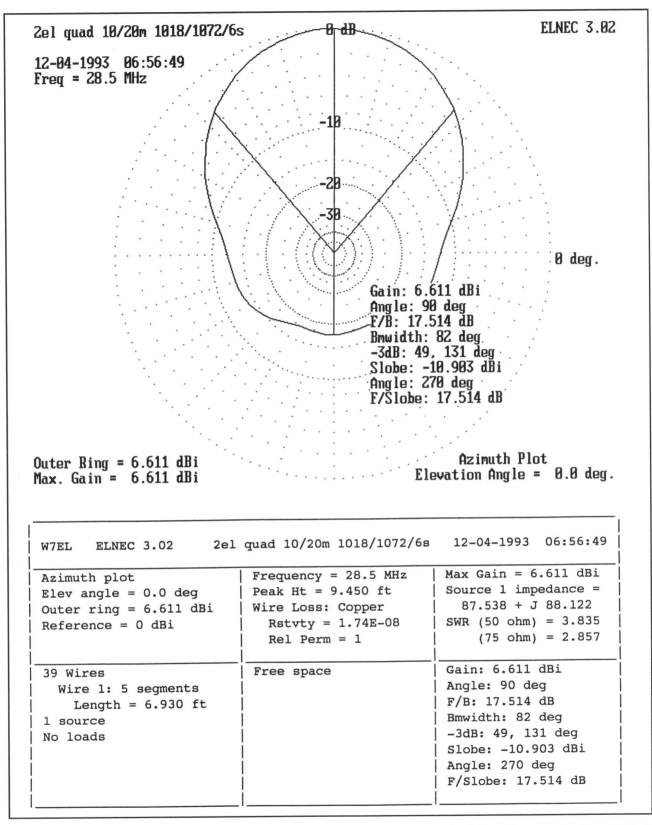

Figure 9-9. Shown here is the 10 meter pattern, with the 20 meter antenna also involved. Roughly, we are looking at 0.5 dB difference in gain from individual feed plus a reduction in front-to-side discrimination. However, one could say the bottom line here is that feeding the two beams together does not create too bad an effect. It is very important, though, to keep in mind that this is only one test. Adding three more bands, as in a five-band quad, would require a great deal of analysis.

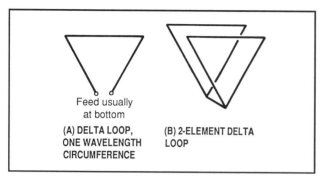

Feed usually
at bottom

**(A) DELTA LOOP,
ONE WAVELENGTH
CIRCUMFERENCE**

**(B) 2-ELEMENT DELTA
LOOP**

Figure 9-10. At (A) is the basic Delta Loop, normally fed at the bottom and used with a gamma match. The impedance without matching is slightly more than 100 ohms. At (B) is a beam configuration. Actual construction details can be found in The ARRL Antenna Handbook.

the quad is fed at the bottom center, which provides horizontal radiation. I didn't mention it above, but there is a slight vertical radiation component from horizontal feed, and vice versa if the feed is vertical. Such radiation, however, is minimal.

Cebik Radiation Plots

After recently publishing an article on quads versus Yagis, a very good antenna engineer, L. B. Cebik, W4RNL, sent me some plots of a program he ran on ELNEC. As he carefully pointed out, the computer programs have some constraints which must be observed. Copies of his plots have been included here with his permission.

For the amateur who may not understand, let me first state the problem. Earlier in this book we discussed resonant antennas. In quads a single full-wave resonant loop is used as the driven element. Now let's take a look at the problem. Let's assume a two-band quad has a 10 meter driven loop with a 20 meter driven loop across it, both connected to a single coax feed. It so happens that 10 and 20 are harmonically related, and in this case a two-wavelength loop is being hung across the 10 meter one-wavelength loop. As I have already pointed out, going back to impedances, the impedances of these two loops will not be too dissimilar. Consequently, and this is what becomes important, power is going to go to both antennas, regardless of the band in use. Both antennas will radiate, and the important question to be answered is how much one antenna pattern changes the other.

Four plots are shown in figures 9-6 through 9-9, 10 meters fed by itself and 10 meters in conjunction with 20 (based on the dual configuration of figure 9-5).

Also shown are 20 meters by itself and 20 meters in conjunction with 10 meters. Without going into a long explanation of the gain numbers that are derived from free-space patterns, note the differences in gain (slight) in either case.

Note that the 20 meter pattern has very slight changes, while those of 10 meters are more pronounced. There are other constraints to be taken into consideration, but these plots certainly show us that there are differences in individual versus common feed.

Delta Loop Beams

I haven't touched on Delta Loop beams, which consist of full-wavelength elements, the same as a quad. The Delta Loop was designed by Harry Habig, K8ANV. In the beam configuration the triangle (see figure 9-10) is fed at the bottom. There are two main radiation points—the bottom where the antenna is fed, and at the center of the top of the triangle. The two sides of the triangle customarily are mounted on the boom with a wire connecting the top. One of the advantages of this method is for a given boom height, additional height is also achieved while providing a lower radiation angle. One of the disadvantages of the Delta Loop is when multibanding, completely separated antennas are required, making a physically "tough" job of boom mounting. Gains of a two-element quad are very similar to those of a two-element Delta.

Monoband Quads

My good friend Paul Carr, N4PC, came up with a monoband quad that is a real revelation. He described the antenna in an article in *CQ* magazine which he has graciously agreed to let me use in this book. The antenna is a reduced size quad that is a very good performer and well worth your consideration.

THE N4PC SQUAD (Squished Quad)
By Paul Carr, N4PC

How would you like to have a quad for 17 meters that is no larger than a 10 meter quad? Sounds great, doesn't it? "But how about performance?" you may ask. Actually, this reduced size beam does very well. The forward gain is down by a fraction of a dB. It is calculated at approximately +5 dBd, and the front-to-back ratio seems to be between 15 and 20 dB. This little antenna will surprise you, and it is rugged, inexpensive, and easy to build.

Background

For many years designers have tried to build small, high-performance antennas by using a variety of electrical load-

This is Paul Carr's "SQUAD" (squished quad). Note how the loop lines are brought in towards the boom. This provides the electrical full wavelength required.

ing techniques. Many of these techniques have provided disappointing results. The most promising technique I encountered during my research was linear loading. This technique has been used by individuals and commercial companies with a variety of results. Perhaps it is time for another try.

I had read of linear loading techniques in the RSGB publication *HF Antennas For All Locations* by L. A. Moxon, G6XN. The next time I encountered linear loading was in December 1992. I worked Andy Pfeiffer, K1KLO, who was using a linear loaded quad. He seemed well pleased with the results. The 17 meter antenna was his second design of this type. (His first was for 12 meters and had been in service for two years.) He promised to send me a copy of his design notes. I was ready to build a version for myself.

In January 1993 I built a single loop (driven element only) and made some preliminary measurements. The impedance was about 50 ohms, and it had the typical figure-eight pattern. Foul weather in January and the blizzard of '93 in March prevented me from building a two-element model. Andy's notes arrived in May, and I was pleased to see that the physical size chosen independently by both of us was about the same. I was sufficiently encouraged to complete my construction project.

Just A Bit of Theory

Almost any antenna book you read will state that the majority of the current in a half-wave dipole is in the center of the wire. Nowhere are "majority" and "center" defined. I did a simple mathematical analysis and found that about 87 percent of the antenna current was contained in the middle 67 percent of the wire. *[In antennas it is customary to*

Another view of the SQUAD beam. With all the tall trees around, one can quickly see why Paul Carr, N4PC, likes wire antennas.

expect maximum radiation from the high current areas of the antenna.—W1ICP] Since a full-size loop is two half-wave wires folded into a square, it seemed logical to make a small loop (60 to 65 percent in size) and fold the ends of the dipoles toward the center of the antenna as linear load-

This close-up of the SQUAD beam shows the neat construction using ordinarily PVC for spreaders. This high-performance beam can be built for just a few dollars.

ing stubs. These stubs are at the low-current points, which helps to minimize loss. Things looked promising, so to the construction phase. This description is for an 18 MHz antenna, but the sizing could be used on any band.

Construction

The choice of spreader material is largely dependent on the builder and geographic location. Choices range from bamboo to fiberglass. I chose schedule 40 PVC (the type intended for hot water). This material was available locally, and I could use standard wood-working tools. Eight spreaders of about 6½ feet in length are required, so I chose 1 inch schedule 40 PVC for the main section of the spreaders and then extended each to the required length by ¾ inch material. These two sizes will not telescope, so I sawed slots of about 3 inches long in the ends of the larger material. After I made the saw slots, I coated the ends of the smaller material with PVC cement and slid them into the larger material. These pieces were held together by small radiator hose clamps until the cement dried. I then removed the clamps. To test the durability of the structure I pounded each spreader on the ground. No spreader broke, so it was time for final assembly.

I attached the spreaders to two identical spiders, made from ¾ inch angle iron, which were welded at the center. I drilled the spiders to accept muffler clamps which attach the spiders to the boom. Again, I used radiator hose clamps to attach the spreaders to the spiders.

I measured 6 feet 4 inches from the center of the spider along the vertical spreaders and drilled a hole for the antenna wire (see figures 9-11 through 9-13). I then measured 6 feet 4 inches along the horizontal spreaders and cut off any extra PVC. I next placed a ¾ inch "TEE" on each horizon-

tal spreader and cemented them in place. This structure was to provided support for one end of the linear loading stub on each horizontal spreader.

I next cut eight pieces of ¾ inch PVC. Each piece was 7 inches in length. I cemented each short piece into the "TEE" on each horizontal spreader. I then measured from the center of the "TEE" along the short pieces of PVC and drilled holes on 4 inch centers (two holes in each short piece of PVC). This provided proper spacing for the linear loading wires.

I next cut four pieces of ¾ inch PVC 14 inches long. I began at the center of each 14 inch piece and drilled four holes on 4 inch centers. These pieces provided the necessary support for the linear loading stubs near the center of the structure.

The wire came next. I used No. 14 stranded plastic insulated wire for all but the stubs, where I used No. 14 solid bare wire. The bare wire made final adjustment of the stubs easier. I left the stubs about 6 inches longer than necessary (a total of 24 inches in each stub), since it was easier to remove wire than to add it.

I then placed the perimeter wire and soldered it to the stub wires, leaving the stub wires near the center twisted together until final proper length would be determined.

As I stated earlier, the impedance of a single loop is about 50 ohms, so I chose to mount the elements on a 10 foot boom. (I was hoping to be able to feed the array directly with 50 ohm coax and eliminate a matching problem.) I used 1½ inch electrical conduit for the boom.

Tests and Measurements

Tuning consisted of adjusting the stubs on the driven element and reflector to provide optimum gain and front-to-

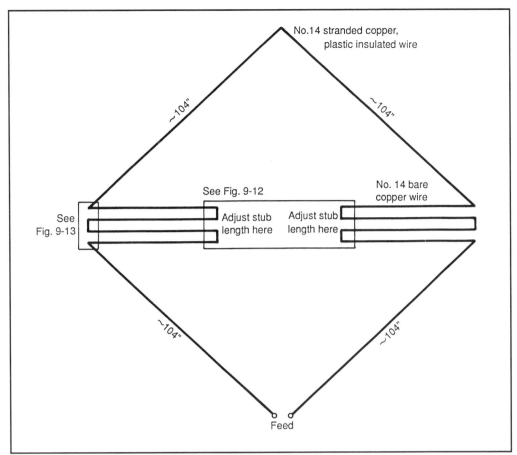

No.14 stranded copper,
plastic insulated wire

~104"   ~104"

See Fig. 9-12

See
Fig. 9-13

Adjust stub
length here

Adjust stub
length here

No. 14 bare
copper wire

~104"   ~104"

Feed

Figure 9-11. This is the basic design of a SQUAD element. As can be seen there are two loading stubs on each side of the element. In actual practice, a simple method of finding the correct stub lengths would be to use an SWR analyzer to determine exact resonance of the driven element. The reflector would be tuned by field strength measurements.

back ratio consistent with an acceptable SWR. I use the word *optimum* because maximum gain and front-to-back ratios are not simultaneous occurrences. A point can be found that will provide good forward gain and a good front-to-back ratio, but experimentation is required. The SWR on my antenna when properly tuned was 1.3 to 1 when fed directly with 50 ohm coax.

Results

My initial contact using the SQUAD antenna was with my friend Lew McCoy, W1ICP. We put the antenna through its paces and checked the front-to-back and front-to-side ratios. We also made gain comparisons with other antennas at my QTH. My second contact occurred when I broke a pile-up and worked Zimbabwe. I knew I had a winner!

Acknowledgements

Good friends are handy when you begin a design and construction project. Both Lew McCoy, W1ICP, and Jim Lindsay, W7ZQ, provided encouragement and instruction. Thanks, gentlemen. Next a multiband version?

The Sommer Beam

Still another concept in multiband beams is the Sommer, DJ2UT, beam. This is a multiband, 20

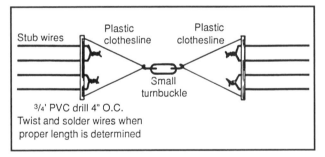

Stub wires   Plastic clothesline   Plastic clothesline

Small turnbuckle

3/4' PVC drill 4" O.C.
Twist and solder wires when proper length is determined

Figure 9-12. Details of the stub ends of the SQUAD.

through 10 meter antenna that has no traps. Frankly, I have always considered this to be an ingenious antenna from several aspects. The antenna essentially is not a Yagi type, because most of the elements are driven. Yagis and quads are parasitic arrays in that only one element is directly fed. In the Sommer array there are basically three primary elements—a full-size, three-element, 20 meter array. These elements are a driven element, a director, and a reflector, all of which are fed. On 20 meters the three elements produce a typical three-element gain of 7-plus dB. On 15 meters, however, these elements essentially become 5/8 wave-

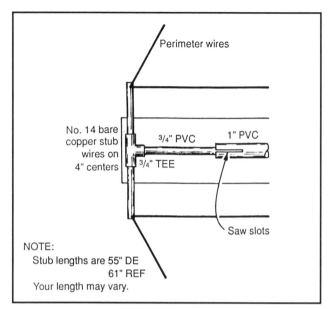

NOTE:
Stub lengths are 55" DE
61" REF
Your length may vary.

Figure 9-13. This shows how the PVC "T" end is arranged for the wire installations.

length long, because the feed gets complicated,, and a 15 meter driven element is also added to the system. On 10 meters the 20 meter elements become full-wavelength elements.

The feed impedance on 10 meters with a full-wavelength linear element will be very high. Therefore, another driven element is added. Because a cloverleaf pattern is generated by full-wave elements, this pattern is corrected by adding more 10 meter elements. I will not go into details of the WARC bands, but they also are added.

What we end up with is a beam that has no trap losses, has reasonably spaced elements, and is a very good performer. On the opposite side, the antenna is heavy (close to 100 pounds or more), and it is expensive when compared to a quad. (However, the same is true of some of the larger trap beams.)

The KØSR Multiband Quad

On the air one day I ran into Steve Root, KØSR, who had a truly outstanding signal. We talked about his antenna system, and I found that he was using a four-band quad—40, 20, 15, and 10 meters. The 40 meter quad uses two reduced size elements with a loading system employing the ON4UN. This reduced quad is quite a bit smaller than a full-size unit. Does it work? Steve has over 300 countries confirmed on 40, which honestly is a tremendous achievement, because his home is in St. Paul, Minnesota. That area of the United States is afflicted by the aurora curtain, and DX on 40 comes hard.

As a construction article, I talked Steve into describing the antenna, plus giving some of his thoughts here. The article appeared in the July 1994 issue of *CQ*. I wouldn't hesitate to recommend this antenna as a 40, 20, 15, and 10 meter array with very respectable gain.

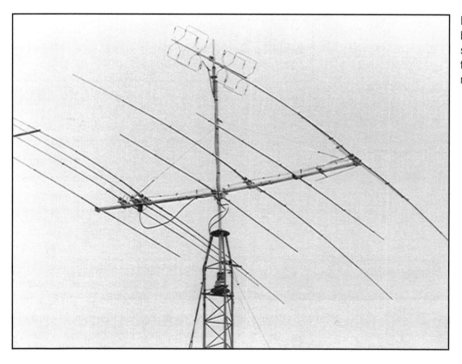

Here is the Sommer, DJ2UT, multiband beam that is being used at my station, W1ICP. Above the beam at the very top is a Swiss quad for 2 meters.

Steve was kind enough to furnish the following material. I might add that the element shortening method he used can be applied to all bands, and the resulting gain for two elements is on the order of slightly over 5 dB with approximately 20 dB front-to-back.

A Compact, Four-Band Quad Array

By Steve Root, KØSR

One of my favorite bands is 40 meters, and while wire antennas are sometimes suitable, they really are not my cup of tea. I wanted a beam, and after much thinking and searching, I decided that a quad was the answer.

My goal was to design an antenna that had gain and directivity on 40 without severely compromising the higher bands. Also, most important, the antenna had to fit my space. After much study, the answer became apparent—a two-element quad on 40 meters. However, a full-size 40 meter quad was just not practical. But a quad, even if it was electrically short, would provide at least 5 dB of gain, plus good front-to-back ratio.

From my studies of quads, I found that with careful attention paid to symmetry, a good, clean pattern would result. Not to be ignored was the fact that if I constructed a two-element 40, it would also provide room for at least three more bands—three elements each on 20, 15, and 10. These additional antennas could push 9 to 10 dB gain each, all fitting on the same boom.

The antenna had to be compatible with my existing tower and rotor and have a total wind area of less than 15 square feet. Also in my design criteria was the antenna could be assembled by one or two people, making the project even more feasible. If an army equipped with a boom truck was required, then it simply was not practical for the average installation. The final consideration was cost. If the project exceeded the cost of a 40 meter Yagi, a tribander, and the associated rotor, then the project would not be practical.

Configuration

Because of the proximity of my tower to the house, a boom length of 18 feet was chosen. This allowed me to use the roof as a platform from which to work. I also favor shorter booms for mechanical reasons: wind loading isn't as severe, so it is easier on the rotor. The 18 foot boom length also works out well for the three-element arrays on the other bands. Therefore, my first consideration was boom length. Because element spacing for 40 meters isn't critical with two elements, spacing with a 40 meter driven element and reflector could be anywhere from 16 to 24 feet.

As I said, a full-size 40 meter quad would have been rather large, so the 40 meter elements aren't full size. The spreaders required for a full-size element 40 meter quad would have been 26 feet long, plus each side at 30 feet plus, and that was simply too big. Techniques for shorten-

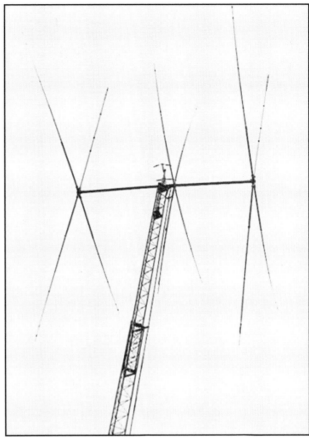

This antenna of Steve Root, KØSR, is a tried and tested design. The idea of working and confirming over 300 countries on 40 meters from the middle of the northernmost part of the lower 48 is a real accomplishment. St. Paul is actually in an aurora area, which makes working DX tougher than in other parts of the country. The addition of 20 through 10 is just icing on the cake.

ing an element have proven successful in other applications, with a lower practical limit being about 70 percent of full size.

For the reader who doesn't understand quads, let me give a little explanation. A regular-size quad uses elements that are approximately one full wavelength in circumference. Also, a quad has four equal-length sides. In addition, a quad can be put up in a "square" configuration or a "diamond" shape (the sides are always equal in length in either configuration; see figure 9-14).

In my case, reducing the size of the 40 meter quad meant that each of the four sides would be 25 feet in length. Thus, 25 feet on a side, with 18 foot spreaders, is what I used. The shortened element is tuned to resonance by adding capacitance at the side corners of the element using additional lengths of wire (see figure 9-15).

This method gives the advantages of being easier to adjust than linear loading, and the loading isn't in the high-

A section of rubber hose is used between the vaulting pole and regular fiberglass.

current part of the loop. The high-current points are where most of the radiation from the element occurs.

Element lengths for the higher bands were developed empirically, based on what I've used in the past. Resonance in my case is biased toward the CW end of the band. Other published dimensions would probably work as well (see *The ARRL Antenna Handbook* and Bill Orr, W6SAI's *All About Cubical Quad Antennas* book).

A quad loop is a lower Q device than a dipole, so its length isn't as critical. The parasitic elements are tuned closer to the driven element for the same reason, typically being only three or four percent different in length.

As mentioned above, there are two possible configurations for the quad—the "square" and the "diamond." The square doesn't hang down the tower as far as a diamond shape, but from a mechanical standpoint the diamond is superior. It lets freezing rain water run off the wires instead of accumulating. With the 40 meter elements it also allows

for longer tuning wires. Also, the current points of the loop are farther apart, increasing gain (gain from stacking).

Most of the wind load is in the big elements at the ends of the boom. A mast-to-boom truss removes the bending movement from the boom and makes the whole array much more rigid. A double truss was used to provide lateral support—one truss on each side of the mast, because a single truss would have the center upright spreader in the way.

Choice of Materials

The boom is a piece of 3 inch diameter aluminum irrigation pipe. This material is readily available, and going to the larger diameter makes for a stronger design.

The spreaders were fabricated from vaulting poles and commercial 13 foot fiberglass spreaders. Fiberglass holds up well to weather and is quite strong. Vaulting poles are readily available and are extremely strong (for source, see the notes at the end of this article). It might be possible to extend standard spreaders with aluminum tubing, but the introduction of other conductors in the near field of the antenna would be undesirable.

The wire in a quad is obviously the actual antenna, but remember that it also is an important structural member as well. When the wind blows, the wire sees considerable flexing and straining. Because of this, stranded No. 14 copperweld was used for the element wires on all quads. I also have had success with a single strand of No. 18 copperweld, which is very inexpensive and light. Do not use soft drawn copper wire because it will stretch forever.

A double truss is used to support the boom and give added strength to the antenna. This Steve Root, KØSR, beam has gone through several severe winters in Minnesota. Also visible are the relay boxes for switching feed lines.

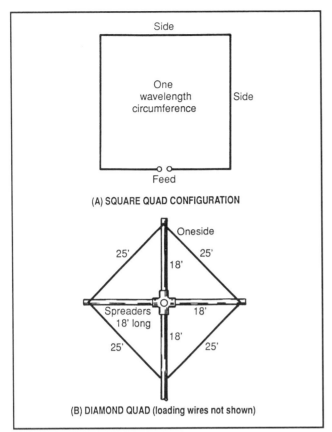

Figure 9-14. There are two quad configurations that can be used—the square (A) and the diamond (B). In my case I used the diamond configuration. As can be seen, for the 40 meter quad my sides are 25 feet, and the vaulting pole spreaders are 18 feet from the boom to the outside. The actual wire dimensions are given in table 9-1. At (B) I have only shown the wires for the 40 meter loop. The other bands fit inside the loop.

The spreader-to-boom clamps are commercial units. I don't know if there is any way to fabricate a clamp that is as strong, light, and straightforward as a one-piece aluminum casting. These clamps have been used for years on various quads I built and have never failed. The boom-to-mast plate is a piece of scrap aluminum and some muffler clamps.

Construction Methods
I used three different-size spreaders. The middle element in the array uses standard 13 foot commercial spreaders. The end elements use 18 foot spreaders made from standard 13 foot spreaders and vaulting poles. The upright spreaders in the diamond configuration hold up most of the weight of the wire, so they have to be stronger than the others.

This antenna uses 15 foot long vaulting poles with a 3 foot extension, cut from a standard spreader. The side and downward spreaders are standard 13 foot spreaders extended with a 6 foot piece of vaulting pole. It broke my heart to

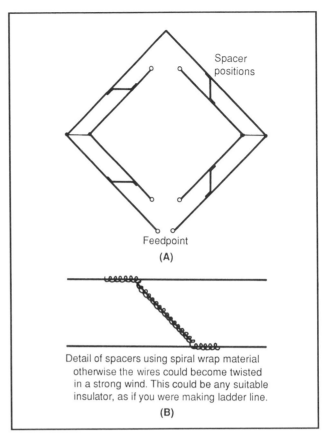

Figure 9-15. (A) shows the addition of wires to the corners of the Root Quad to provide the capacitance required to tune the shortened elements. At (B) is the method used to maintain the required separation.

take a saw to those beautiful 15 foot vaulting poles, but it is less expensive to use standard spreaders to make up the length, and you don't need the full diameter out at the end with the light-duty spreaders. This minimizes wind load.

To join the standard spreader to the vaulting pole, a shim was made from a section of black rubber hose found in the local hardware store. Although close to the proper wall thickness, etc., it required some sanding to achieve a snug fit to the inside diameter of the vaulting pole. A 6 inch long slit was cut lengthwise in the end of the vaulting pole, and a stainless-steel hose clamp was used to make a compression fit. No slippage has been observed.

The wires were cut to length using a 100 foot tape. A 1 inch allowance was made on both ends for splicing. If the splice in the wire lands on top of a spreader, considerable flexing may occur at the solder joint. For this reason, the splice should be offset. Then it will see only a pulling stress when the spreaders move in the wind.

To attach the wire to the spreader, I prefer the method first described by Landskov.[1] The wire is first tied to the spreader using string in a criss-cross fashion. This is followed by fiberglass filament tape and then regular electrical tape. The filament tape will deteriorate due to ultraviolet radiation, but the

Band	Reflector	Driven	Director (ft.)
40	100	100	
20	73	71	68.375
15	49.083	47.5	46.251
10	36.542	35.19	34.417

Table 9-1. KØSR's element lengths for the four-band quad array.

electrical tape will shield it. A cable tie over the tape will keep it from unraveling into little banners flapping in the breeze. This method allows for some adjustment of tension, and also eliminates the need to drill holes through the spreader. The traditional method, using holes, will weaken the spreader and may result in breakage. Keep in mind that my antenna has been up for a few years in severe Minnesota winters, so I know the techniques prove out.

To maintain proper current distribution in the element, it is a good idea to keep the assembly as square as possible. To do this, the wire is marked off with pieces of tape at each "corner" position. The attachment points out on each spreader were marked with a pen. The two points should intersect; if they don't, something isn't square. The dimensions I used can be found in table 9-1. Start with the smallest loop (10 meters) and work outward. When the 40 meter elements are built, the tuning wires should be made as long as possible, since they will be trimmed later to tune the element to the proper resonant frequency.

It isn't necessary to string the element like a violin! Even though this may appear pleasing, the extra tension will promote wire breakage. Just enough tension should be used to hold the element in shape. The spreaders should flex. Think about the abuse a fiberglass fishing rod can handle!

The feed point of each driven element deserves special consideration. Small loops are formed at each end of the wire, and a piece of string is run through these loops and around the spreader to make the attachment. The filament tape and electrical tape are applied over the string to complete the mechanical attachment. Next the coax is soldered in place, and the whole thing is weatherproofed with a product such as Coax-seal®. It is very important to keep moisture out of the end of the coax. With the diamond configuration, the downward spreader provides a solid position for the feed point, and the feed line can be dressed along the spreader and back to the boom. This will minimize movement at the solder connections.

The spreaders are held to the boom clamps with stainless-steel hose clamps. It isn't necessary to "crunch down" on the spreader; just hold it firmly in place. Otherwise a broken spreader can result.

Each element is assembled on the ground and brought up to the boom one at a time. I have a crank-up tower, and when it is nested, I can reach the end of the boom from the roof of my house. Alternatively, a tall step ladder could be used. With a free-standing tower, the boom could be placed at an appropriate height for assembly and then the completed antenna hoisted up to the top.

Tuning and Feeding

Proper current distribution is necessary for a good antenna pattern. This is achieved with element geometry and with a good balun at the feed point. The current-type balun works well and is compact. It consists of a number of ferrite beads placed over the outside of the feed line, near the feed point. The amount of inductive reactance is proportional to the number of beads and frequency. I used "Super Beads," available from Radiokit, Amidon, and The Wireman. Radiokit also sells a balun kit that uses 5 beads and is rated down to 160 meters. I used five beads on 40 meters, four beads on 20 meters, and three beads on 15 and 10 meters. The beads were taped up and sealed against moisture; otherwise they could crack in the winter.

There are two alternatives to the beads. One is to use a linear sleeve balun. This device is made from a quarter-wave length of braid placed over the feed line and electrically attached to the feed-line braid at the end away from the feed point. It is inexpensive and simple to construct.

Another simple method of keeping the feed coax shield "cold" for RF is to grid dip the shield with the antenna attached. If a dip shows up in the band, simply add a few feet of coax to get the dip out of the band, and the shield will be cold, preventing feed-line radiation.

A separate feed line for each band is run to a remote antenna switch placed on the mast. Even though tying all the feed points together has been a standard practice, I know that common feed will cause interaction problems, both on 40 and 15 and 20 and 10, complicating impedance matching. There are commercial units available, or one could be constructed using relays.

The loaded 40 meter elements are tuned to resonance by introducing capacitance at the side corners (see figure 9-15). This method was developed by G3FPQ, and was described by Devoldere in his book.[2] The technique has been applied differently here due to the diamond shape of the element. Longer wires are possible this way.

The spacers seen in the drawing keep the tuning wires from twisting around the element during windy conditions. Any light-insulating material would work. I used pieces of "spiral wrap" material (used to bundle up wires) with a stick slid down the middle. The ends simply were wrapped around the wires.

The four-band reflector was built first and placed on the tower. The lower corner of the 40 meter element was made like a feed point. This permitted a small loop to be placed in the loop so that the resonant frequency of the element could be measured. This was done using two different instruments—a grid-dip meter and a noise bridge.

The length of *all four* tuning wires is trimmed to keep them the same length and to maintain proper current distribution. The reflector was tuned to 6.8 MHz, or 3.5 percent below the design frequency. The driven element was then tuned for the best impedance match at 7.05 MHz. This may not be the best way to tune a parasitic array, but it worked! The impedance match of the higher bands is primarily determined by reflector spacing, which in this case is fairly close to 50 ohms. The addition of a matching device is undesirable due to the need to weatherproof it, and adjustments are difficult up in the air. It really is unnecessary.

Performance

It is difficult to estimate how well an antenna works in absolute terms. The previous antenna used on 40 meters was a quarter-wave vertical with an elevated ground system, and it was left in place for comparison purposes. On both stateside and DX signals one to two S-units improvement with the quad was typical. In no case was the vertical better than the quad, and due to the low noise characteristic of loop antennas, it was surprising how often it was possible to copy signals that simply weren't there with the vertical.

My quad is on a 50 foot tower, which is really quite low in terms of a wavelength on 40 meters, yet the antenna exhibits a reasonable pattern. It has been observed that in DX pile-ups this array will outperform vertical or wire arrays, and hold its own with Yagis at 70 to 100 feet.

An antenna that doesn't stay up isn't worth much, so most of the design centered on mechanical considerations. This quad has been up since August 1990 and has survived Minnesota winters and nasty thunderstorms without a problem. How good is this shortened quad on 40? I have over 300 countries confirmed on 40—no small feat for this area of the United States.

Appendix

Suggested additional reading: *All About Cubical Quads,* by Bill Orr, W6SAI, 2nd edition, p. 46; *ARRL Antenna Anthology,* 1st edition, p. 56; *ARRL Antenna Handbook,* 15th edition, p. 12-2.

For vaulting poles contact Peterson Co., P.O. Box 25536, Salt Lake City, UT 84125 (801-972-3328).

For quad components contact the following:
Cubex Company, P.O. Box 732, Altadena, CA 91001.
Antenna Mart, P.O. Box 699, Logansville, GA 30249 (404-466-4353).
Lightning Bolt Antennas, RD #2, Rt. 19, Dept. Q, Volante, PA 16156 (415-530-7396).

Additional Reference Material

I said earlier that I was not going to give any more formulas. Well, I lied. Over the years the formulas for quad elements have more or less been standardized. The following formula should be among your reference material. For the driven element use 1005 divided by the frequency in MHz, for the reflector use 1030/F(MHz), and for the director or directors use 975/F(MHz).

In the further reading department are the following publications.

One of the best quad men around today is James "Doc" Lindsay, W7ZQ. Lindsay wrote an article which appeared in the May 1968 issue of *QST*.[3] The article presented design details of a quad, including dimensions which have become more or less the standard. An excellent book on quads by Bill Orr, W6SAI, is *The Quad Handbook.* Another good reference and guide is *The Quad Antenna* by Bob Haviland, W4MB, published by *CQ.* For the details required to make a quad with more than two elements, read the "Quad Array" chapter of *The ARRL Antenna Book.*

Footnotes

1. H. Landskov, "Evolution of a Quad Array," *QST,* March 1977, p. 32.

2. J. Devoldere, *Antennas and Techniques for Low Band DXing,* ARRL, Newington, Connecticut.

3. J. Lindsay, W7ZQ, "Quads," *QST,* May 1968.

Chapter 10

One-Element Rotary Dipoles

Many amateurs miss a real opportunity when they overlook the fact that rotary dipoles make excellent directional antennas. Quite a few years ago I described an inexpensive rotary dipole for 15 meters that offered a perfect match to 50 ohm cable. This antenna enjoyed great popularity because the entire antenna (at that time) could be built for less than $10. (It still is very inexpensive, as you will see.)

Let's look at a dipole technically, and then consider it as a rotary. As I pointed out, the impedance of a half-wavelength dipole is on the order of 50 to 70 ohms, depending on its height above ground. The radiation pattern of this antenna is a figure-8, with two maximum lobes broadside to the plane of the antenna. From what you have read and learned in basic theory so far in this book you know one important fact about gain—-that is, the manufacturers of transceivers calculate or set one S unit as 6 dB. Let's try to clarify this widely used statement.

The statement that one S unit equals 6 dB may or may not be true. It is extremely difficult or next to impossible to make an S meter that will indicate exactly 6 dB for every S unit.

Components including meters will vary from receiver to receiver. The only way to know for sure how your receiver is calibrated is to use a very accurate signal generator and actually measure the S-meter reading in microvolts. However, in this discussion let's assume that the S meter does equal 6 dB per S unit.

Dipole Performance

When we rate antennas, we compare their gain to a standard dipole, the half-wavelength type I mentioned previously. The radiation pattern of a dipole, in its

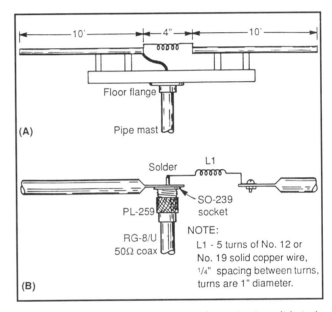

Figure 10-1. (A) My 15 meter, one-element rotary. It is truly amazing how good a performer this antenna can be. (B) The details for the reactance cancellation inductor.

best direction, is down just about 6 dB as compared to the ordinary beam. What does this mean from a practical standpoint?

At this point let's do some comparisons. Let's say that Joe Ham down the street from us is using a trap beam, and we have a rotary dipole. Let's also say we both contact another local, Pete, about a mile away. Both our signal and Joe's signal, for the sake of discussion, will be ground wave—no sky wave involved. We can expect, and it will be true, that our signal will be exactly one S unit weaker than Joe's as read on Pete's S meter, assuming that one S unit equals 6 dB.

Now suppose we start reaching out for contacts via

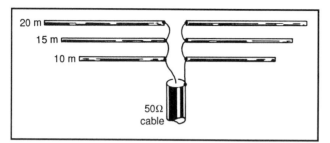

Figure 10-2. The electrical drawing of the three rotary dipoles described in the text. The feed connections at the dipole should be kept as short as possible.

sky wave. Believe it or not, considering the power gain of Joe's beam, he may or may not be stronger than us at a given distance! This is strictly because of the sky-wave angles involved and ionosphere reflection factors. To be honest, Joe normally is going to do better than us, but not all of the time, and that is what is important. Naturally, a rotary dipole is not going to have any front to back, as the front is the same as the back. However, it does have good front-to-side rejection, not a great deal different from a beam. The fact that the dipole has no front to back is, in one sense, an advantage. Keep in mind that we only have to rotate the dipole 180 degrees—not 360 degrees—to get the best results.

Construction Details

Figure 10-1 shows the construction details of a rotary for 15 meters. Basically, the antenna consists of two lengths of $1/2$ inch diameter electrician's thin-wall conduit, each one 10 feet long. These sell for just a few dollars a length. (Hard-drawn copper tubing could be used, but that gets expensive.) One of the ends is flattened in a vice so that it can be drilled to take an SO-239 coax chassis fitting. The normal length of a 21.2 MHz dipole is approximately 22 feet, meaning that the overall 20 feet is about 10 percent short. Also, as you learned at the beginning of the book, a short antenna's feed point is capacitive reactive, which means a slight of amount of inductance must be added to cancel out the reactance. It so happens in this particular case that we are left with exactly 50 ohms of impedance, a perfect match for 50 ohm coax, and that is what we use. The coil is connected

from the SO-239 inner pin to the other half of the dipole. The coil can be constructed from $1/4$ inch diameter copper tubing (used in refrigerator work) or a length of No. 10 or No. 8 solid copper wire. The object here is to use a coil material that will hold its shape without being mounted on a form.

To support the antenna I mounted it on a length of 2×2 wood and made standoff insulators from PVC. The wood then had a pipe floor flange attached, and of course pipe for a mast. As I stated, the entire system is very inexpensive, but very, very effective as a monoband, one-element beam.

Adding Additional Bands

For the other bands, the thin wall can simply be extended to make a dipole applying the formula 468/F. If necessary, the 10 foot thin wall lengths can be extended using stiff wire mounted on the ends of the thin wall to achieve the half wavelengths. I flatten the outside ends, and about one half inch from the end I mount a nut and a screw to hold the wire extenders (coat hangers have good "springy" wire useful as extenders).

For properly calculated dipole lengths, and with 50 ohm feed, the match should be close enough for all modern transceivers.

While I have never tried it, it might be worthwhile to mount three dipoles, one for 10, 12, and 15 on a single 2x2; one on top; and one on each side. Support standoffs could easily be made from pieces of PVC. The feed points of the three dipoles would be handled as shown in figure 10-2. The inner conductor of the coax connects to the three halves of one set, and the outer braid connects to the other three halves. The dipole halves at the feed ends should be kept no more than 1 inch apart.

All three feed points are not inclined to "see" each other electrically because the reactances on the unused bands would be very high, and power would only flow to the band in use. This could prove to be a very low-cost one-element rotary beam for three bands—and, I might add, a worthwhile performer. Keep in mind that with this system, on a strict measurement basis, you would only expect to be down only slightly from the big guns. If you build this system, please let me know.

Chapter 11

Some VHF Antennas

These days nearly every new amateur starts out on VHF, usually 2 meters, so a discussion of popular 2 meter antennas is in order. I described a very inexpensive (coat hanger) mobile vertical antenna a little earlier. For repeater work and packet, DX clusters, and so on the polarization of the radiated signal is almost exclusively vertical. Over years of experimenting, amateurs have settled on vertical polarization for mobile repeater work. That is not to say that vertical polarization is exclusive. Weak-signal, moon-bounce, and a great deal of satellite work is done using horizontal polarization. In our case, however, let's stay with vertical polarization.

Rubber Duckies

Probably the worst antennas for practical communications work are the "rubber duckies" used on hand-held transceivers. Electrically, this type of antenna is a quarter wavelength of flexible wire wound into a

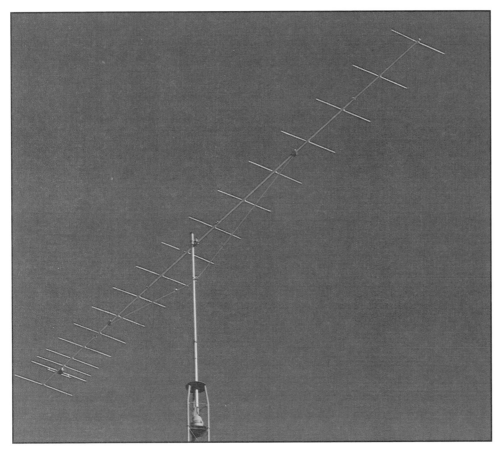

On 2 meters and throughout VHF it is possible to build or buy some very "big" antennas—big in the sense that they have lots of elements. Here is an example of the Cushcraft very high gain 2 meter beam used by many amateurs for serious 2 meter work.

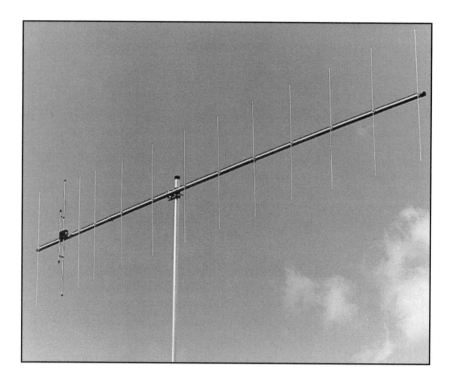

As I mention in the text, in some instances it may be very difficult to access a repeater. One real answer to the problem is high gain, very directional beams. Note this Cushcraft 13B2 is mounted so that its elements are vertical for 2 meter vertical polarization (repeaters). As higher and higher gain figures, with many directors, are reached, the beamwidth of the antenna becomes narrower and requires careful aiming.

short form, usually about 6 inches long. A full-size quarter-wave would be 19 inches long, but some of the stub "duckies" are as short as an inch or so! Is that bad? Yes and no, as you will see.

Repeaters

When repeater stations provided the means to extend the range on 2 meter communications, amateur radio received a real shot in the arm. The VHF and UHF normal range for communications is line of sight. Or, in amateur radio parlance, "If you can't see 'em, you can't work 'em." By installing repeaters in high locations, amateurs started working extremely long distances. This was still "ground wave," or really line of sight. Previously, where one amateur was lucky to talk to another amateur, say, a mile or two away, on 2 meters these distances jumped to hundreds of miles.

The ZIA Network

As an example, in New Mexico, where I live, we have a system called the ZIA network. (ZIA is the indian name for New Mexico.) The ZIA system is based on one repeater relaying its received signal to other

repeaters, which in turn do the same. All of these repeaters are on mountaintops, most at over 9000 feet elevation.

My home is in the southwest quadrant of New Mexico in Silver City. With my hand-held running a watt or so I can acquire a repeater that is on a mountain called Jack's Peak, which is about 3000 feet higher than my location (6300 feet). Jack's Peak is some 30 miles or so from my house. From Jack's Peak my signal reaches out as much as 100 miles in many directions. In addition, the Jack's Peak repeater reaches other repeaters, relaying my signal even greater distances, such as to Tucson, 250 miles away, to Phoenix, close to 300 miles away, or even clear over to California and north to Colorado. Keep in mind that this is all done with a rubber ducky antenna! But, and this is important, the very nature of a repeater being high and accessible is the answer to being able to contact it with a poor antenna.

Unfortunately, everyone is not within easy range of a repeater, and therefore good communication requires a much better antenna. The simplest better antenna is a 19 inch quarter-wave whip with a ground plane (say, three 19 inches radials). Reverting to this antenna may make you solid into a repeater you couldn't reach before. Also, what is important to reach repeaters on this band is the height of your antenna. Every foot in height increases your horizon.

The next step up is a $^5/8$-wavelength vertical. This antenna has more gain than a quarter wavelength and is the most common type of vertical used in mobile work.

Earlier I mentioned Jim Steven, KK7C, and some of his unusual antennas. He has an antenna called the Pico J, which is a full half-wave vertical, completely flexible. This antenna has a matching system at the bottom to provide a good match to 50 ohm cable. The entire system is very flexible. In fact, it comes rolled up and stored in a small pouch. The antenna is fed at one end with 72 inches of flexible coax terminated in a gold-plated BNC fitting to use with your hand-held.

I had not realized just how bad a rubber ducky could be. I was attending a Quarter Century Wireless Association board meeting, and the motel at which I stayed was not the best location for the local repeater. In fact, the rubber ducky was strictly hit or miss—mostly miss. I hooked up the Pico J, and wham, I was full quieting into the repeater. Since then I have been in many such situations, and the antenna has always pulled me through.

It is no big deal to make your own quarter-wave whip and mount it in a BNC fitting. Any stiff wire (the 19 inch length of coat-hanger wire described earlier is excellent) can be used.

You are going to be surprised if you have only used a rubber ducky and haven't tried a good antenna.

Most of these antennas are for mobile work. When discussing fixed locations where you need signal improvement, then we start to think of beams.

Another very simple change that will improve your hand-held performance immensely is the addition of a ground wire, or more accurately, completing the lower half of a rubber ducky. The rubber ducky usually consists of about 19 inches of wire wound around a form, usually 4 or 5 inches long, and then covered with a protective flexible coating. In other words, it is a shortened $^1/4$-wavelength whip without a ground plane to "work" against.

The hand-held receiver has miscellaneous metal in it, but not enough to provide a really a good ground. Adding a ground is very simple and does not require any changes in the hand-held or its antenna. Simply take 20 inches of flexible, insulated wire and remove the insulation from 1 inch of the wire at one end. Remove the rubber ducky, wrap the bare wire around the coax fitting, and reinstall the rubber ducky. A ground—or if you want to call it a half-wave center-fed dipole you may—has now been added. It isn't exactly that, but it will provide tremendous improvement. Try it. You will be pleasantly surprised.

Two Meter Beams

When discussing antennas for 2 meters, gain is very important to achieve your goals. There are many good beams available. For really tough situations, a good

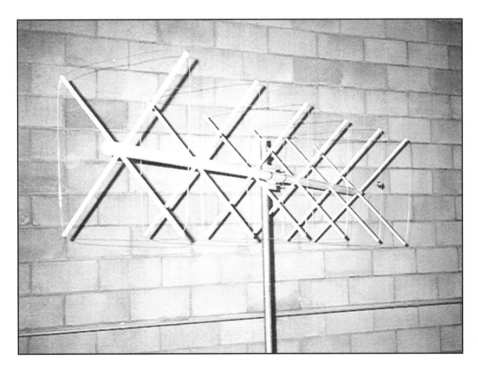

This is a very high gain six-element quad. This model is also relatively inexpensive and is manufactured by Lightning Bolt Antennas. Lightning Bolt makes quads for any configuration and to order.

example is the Cushcraft 14-element Yagi, which has 12 to 14 dB gain (see the accompanying photo). While the gain figures of manufactured antennas tend to emphasize transmitted signal strength in a given direction, it is extremely important to think in terms of receiving gain. Having a better antenna to transmit to a repeater is not nearly as important as being able to *hear* the repeater. Also, while you don't hear the following expression much anymore, it nevertheless holds true. Some repeaters are known as *alligators*—all mouth, no ears. We need a good loud signal coming and going.

I have shown a few different VHF antennas. Multi-element quads are good antennas for 2 meters, and while they do not have the sharp pattern that a 13- or 14-element Yagi produces, they do put out a respectable signal and are very light weight. When quads are fed at the bottom (or top), the polarization is horizontal. For repeater work they must be vertically polarized, so they are fed at the side.

Chapter 12

Multiband Mobile Antennas

A great deal has been written about what is the best mobile antenna. As I pointed out at the beginning of this book, the radiation resistance (Rr) of an antenna depends to a large extent on its length. As the antenna is made shorter, the Rr gets lower. In the case of an 80 meter mobile whip—say, 8 feet overall—we are looking at a very small fraction of an ohm as the radiation resistance. This in turn means that most of the power—in fact nearly all of it—is dissipated in the ohmic portion of the impedance instead of being radiated. How can we overcome this bad ratio of ohmic to radiation resistance loss? The only way I know is to make sure a vehicle has good ground connections. And this doesn't merely mean a good ground at the battery.

The Importance of Good Grounds

Often I recall an amateur who reported into our 80 meter phone net, always when mobile. His signal was very strong—no, make that *extremely* strong!

We all marveled at this particular individual's signal, but he was very close-mouthed when asked about his antenna. At a local hamfest another amateur happened to look under this person's vehicle, and he discovered that every part of the frame and other metal parts were "strapped" together with grounding braid. Some of us then looked under the hood and found the same thing—every metal piece carefully bonded together.

When what we were doing became obvious, the reticent amateur came clean and explained his installation. He knew he had to get that ratio of ohmic to radiation resistance down, so he made sure everything had a good bond. And it obviously worked! The ohmic-loss ratio to radiation resistance was kept as low as possible.

In modern vehicles, with so much non-metallic

The High Sierra mobile/RV antenna was derived from the Don Johnson, W6AAQ, design for the DK-3 antenna. If you are going to be mobile, I recommend Don's book, 40 Plus 5 Years of Mobileering. Here Don is demonstrating the construction methods of his DK-3. All antenna parts are made from readily available materials.

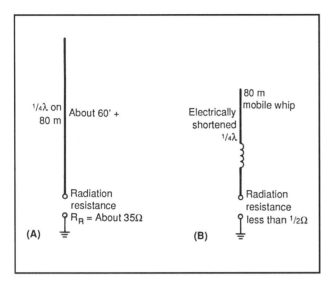

Figure 12-1. Radiation resistance of low-band verticals, 80 or 40 meters, one-quarter wavelength in length, will be on the order of 35 ohms. However, when the antennas are shortened physically, such as in an 8 foot mobile antenna on 80 meters, the radiation resistance can drop to a mere fraction of an ohm. Assuming a ground resistance of, say, 3 ohms, one can see the ground loss ratio is very low in the case of the full-size quarter wavelength. A ratio of approximately 3 ohms to 35 ohms (radiation resistance) is less than one tenth of the power lost. However, in an 80 meter mobile antenna with a radiation resistance of one tenth of an ohm and a ground loss of 3 ohms, the loss ratio is very high, with the majority of the power dissipated in ground. This is why in modern vehicles with partial frames, insulated motor mounts, etc., it pays to do a lot of strap bonding to reduce these ground losses and improve the radiation ratio.

material used, it becomes extremely important to bond all metal assemblies such as motor, frame, etc., together (see figure 12-1). Metal-braid strapping is available from many suppliers.

Assuming that all the grounding has been done, what is the best antenna? We will concentrate on 80 and 40 meters, but the theory holds true for other bands. Argument number 1: What is best base-loading, continuous spiral wind from top to bottom, or center loading? From my experience, and the experience of others who have spent a lot of time testing mobile antennas, probably the worst performer is base loading. Next comes the spiral wrap. Probably the best, or at least the most practical, is center loading. The performance of center loading can be improved if a large, circular capacitive hat mounted near the top of the antenna is used. However, the improvement is not

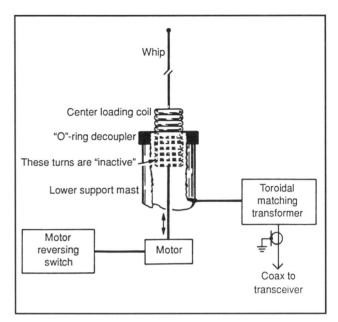

Figure 12-2. Details of the High Sierra vertical, multiband antenna. The 12 volt DC motor is mounted inside the casing of the antenna. The motor drives the coil up and down past an "O" ring, which shorts out unused portions. It is always possible to obtain a 1 to 1 (or extremely close to 1 to 1) match.

really worth the problem of driving with a large circular disk on the top of a whip antenna!

Don Johnson, W6AAQ

To me, the person most knowledgeable about HF mobiling, the guru of this subject, is Don Johnson, W6AAQ. Don has spent over 40 years working on this subject, so he definitely knows what he is doing. Don doesn't manufacture antennas, but he is the author of a book that is a compilation of his work and the work of others.

Literally hundreds of amateurs have built Don's antenna and swear by it. In fact, one amateur lived in a development that did not permit antennas. He bought two older automobiles, parked them in strategic positions in his yard, and installed two of Don's antennas, one on each vehicle. By using switched coax phasing line brought into his shack, he phased the antenna radiation patterns. It only took him a few weeks to work 100 countries! Now that's ingenuity!

For amateurs with limited space—such as condominium owners, RVers, apartment dwellers, and so on—I strongly recommend this antenna as a low-band, 80 through 10 meter installation. Years of experiment-

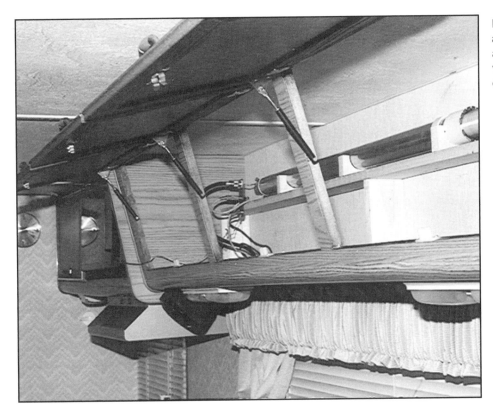

KN6UI mounted his DK-3 inside a cupboard in his motor home and connected one end to a whip mounted outside! He claims it really works very well.

ing have gone into the design. I know it really works and provides excellent multiband coverage.

So what is this marvelous antenna? First, it is of the center-loaded variety. However, Don installed a motor-driven shorting system to short out unused coil turns and resonate the antenna. The motor used is taken from a 12 volt electric drill and is installed inside the antenna—that's right, inside the antenna. From the driver's seat, or from the shack, the motor is merely turned on via a 12 volt DC line and the antenna tunes itself. By observing an SWR bridge at the operating position, the SWR drop to a 1 to 1 match can be seen, indicating resonance.

Installations

Don furnished me with photographs of different installations his friends are using. Some of these are included in this chapter. These will serve to give you plenty of ideas on how to make your own installation.

There are several more basic mobile or RV installation points. To obtain the best signal the loading coil should be above any surrounding metal. Many of the mobile installations show the antenna mounted as high as possible. Good judgment must be used here. I always remember a friend of mine who didn't use good judgment. He mounted an 8 foot whip on the

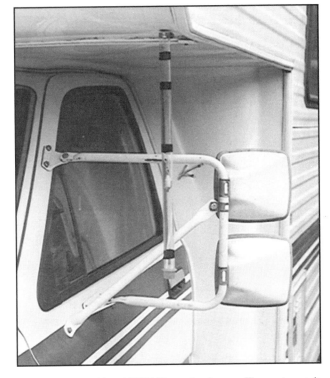

This is the outside of KN6UI's motor home. The antenna is inside and the top whip is outside. He mounted a poly tube on his outside mirror with a fish sinker connected to a line going to the antenna. When he operates the motor to tune the antenna (from his front position), the sinker shows him where the antenna is tuned!

Here is the High Sierra multiband vertical mounted on a vehicle. Note this installation does not require making any holes in the car. Cables are under the trunk lid.

This is the High Sierra showing the coil, which rotates and rides up or down. The quick-disconnect whip is at the top.

front roof of his automobile, resulting in the antenna sticking up 8 feet above the roof. He had a very good signal, but unfortunately one day the tip of the antenna caught on an overpass while he was doing about 50 mph. He tore a metal strip about 5 inches wide from the front to the back of the roof—enough said! Also, while it isn't always practical, try to keep the coil as far away from the metal body as possible.

Don's book is called *40 Plus 5 Years of Mobileering.* If you want to be a real guru, read this book, which is available from Radio Bookstore, Rindge, New Hampshire. The antenna is known as the DK-3. Don provides a shopping list for the parts for this fun project. Similar commercial antennas in this class run over $200. While I don't have exact costs, my guess would be about $70 for the parts, assuming you have no junk box. Amateurs, however, are noted for their ingenuity, and a project like this is also a lot of fun. If the idea of winding 150 turns of wire onto a PVC form to make the loading coil frightens you, don't worry. Don's methods are spelled out clearly enough for a ten-year-old to do the job.

As I stated, I have included photos of different installations using Don's basic DK3 antenna. Note that some of these photos show the antenna coil and

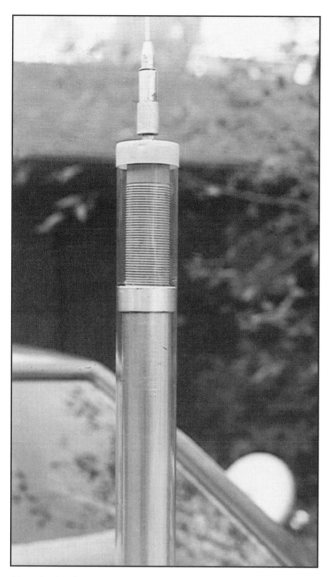

Here is the "rain hat" which protects the High Sierra antenna from precipitation. The hat rides up and down with the coil movement.

The ash tray was removed from the vehicle and the controlling switch for the High Sierra was installed. It provides easy control access from the driver's seat.

The mobile setup of Woody Binford, W6LHH, Novato, California.

tuning device mounted *inside* the trailer, RV, or truck. The "top" whip is mounted outside. Don has told me that this method both protects and hides the main part of the antenna. What is important, though, is that it does not seem to hurt performance.

Incidentally, a commercial version of this DK-3 multiband is the High Sierra antenna shown here. The commercial version is extremely well made, and I would not hesitate to recommend it. Figure 12-2 gives details of the High Sierra unit.

Summary

Well, that just about winds up this antenna book. I have attempted to pass on to the reader those things

Bryan Graham, W5VXV, of Texas City, Texas, ended up with an insulated copper pipe, so he had to hook a piece of braid to connect the pipe to the antenna.

This mobile setup belongs to Patty Smith, WB6DRG, of Torrance, California. Her husband, Frank, is not a ham, but he built this using hand tools in his shop.

John Bergman, N7FGF, of Springfield, Oregon, thought that big would be better, but he found out otherwise. ➔

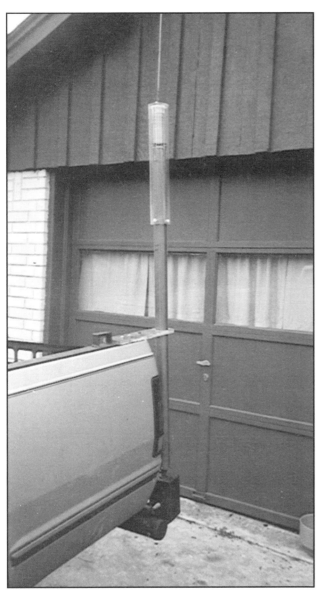

Roger Mundinger, WD5GTQ, of Houston, Texas, made this antenna. Note the coil point marks, which can be seen in the rear-view mirror

This is the mobile setup on the car of WD5GTQ's XYL. The entire mount can be pulled off.

The bumper-mount setup of Bernie Bernardini, K6UV, San Diego, California

In this case the metal bar which holds the antenna mount is attached to the RV ladder with toggle clamps. When Vern Schumann, KN6UI (Los Osos, California), is parked for long periods of time, he raises the assembly to the very top, as shown here. He says he knows of others who have attached the antenna base to a motor (antenna rotor, auto window motor) so they can raise, lower, or tilt the antenna as circumstances warrant.

The simple but rugged mount for the Don Johnson mobile antenna.

The mobile mount is installed on the spare-wheel carrier in this installation. The bottom of the antenna should be as high as possible to keep the radiation in the clear. ➡

that have worked for me over my long career in amateur radio. Most of my time in amateur radio has been spent answering questions. At one time I was the "official answerer" of technical questions for the ARRL. It was called the Technical Information Service. Believe it or not, I totalled nearly 5000 inquiries a year, many of which were about antennas.

Already I can see things that I left out of this book and other antennas that are worth describing. Maybe

we will come out with a second edition!

Again, my thanks to the hard-working gang at *CQ* magazine, and to anyone else who knowingly or unknowingly contributed to this book.

My parting comment for now: Remember that anything that conducts RF is an antenna, be it a paper clip or a large rain gutter. The trick is to get power into such things, and you should know how to do that by now!

Index